ACPL ITEM
DISCARDED
Y0-AAD-674

PREVENT BURGLARY!

An Aggressive Approach to Total Home Security

Mick Davis

Prentice-Hall, Inc.
Englewood Cliffs, New Jersey

Prentice-Hall International, Inc., *London*
Prentice-Hall of Australia, Pty. Ltd., *Sydney*
Prentice-Hall Canada Inc., *Toronto*
Prentice-Hall of India Private Ltd., *New Delhi*
Prentice-Hall of Japan, Inc., *Tokyo*
Prentice-Hall of Southeast Asia Pte. Ltd., *Singapore*
Whitehall Books, Ltd., *Wellington, New Zealand*
Editora Prentice-Hall do Brasil, Ltda., *Rio de Janeiro*
Prentice-Hall Hispanoamericana, S. A., *Mexico*

This publication is designed to provide accurate and authoritative information in regard to the subject matter covered. It is sold with the understanding that the publisher is not engaged in rendering legal, accounting, or other professional service. If legal advice or other expert assistance is required, the services of a competent professional person should be sought.

. . . From the Declaration of Principles jointly adopted by a Committee of the American Bar Association and a Committee of Publishers and Associations.

Library of Congress Cataloging-in-Publication Data

Davis, Mick
Prevent burglary!

Includes index.
1. Dwellings—Security measures. 2. Burglary protection.
I. Title.
TH9745.D85D38 1986 643'.16 85-17014

ISBN 0-13-699265-X
ISBN 0-13-699257-9 {PBK}

Printed in the United States of America

To L. Ron Hubbard and Carole

INTRODUCTION

My interest in home security was inspired about ten years ago by an incident that could have caused my death had I arrived home a few minutes later. As I arrived at my floor, two well-dressed middle-aged men in the hall asked me if I was getting out of the elevator there. I said yes, and as I did, they entered and left. Though I had never seen them before, I was not suspicious as they "looked honest." I went to open my apartment door, near which one of the men had been standing. The lower lock was open and the upper one was almost so, my key able to turn it when only halfway in. These were the expert lock pickers who had been "hitting" the building recently. Had I returned a few minutes later and confronted them in the apartment I might never have lived to write this.

It would be a far better world if there were no need for a book of this kind. Man is a master predator upon his own kind; the worst among us are more vicious than any other species. We all seek safety and security for ourselves, our families, and our property, and it is in these interests that this book was written; it focuses on the prevention of a very common, usually nonviolent crime, but one that is always costly and emotionally traumatic to its victims. Knowledge properly applied can bring security and peace of mind. Use the knowledge given herein and your chances of becoming a burglary victim will be greatly reduced.

Much that has been written on burglary prevention has been of greater value to the criminal than to the victims. Much misinformation has been published which at best gives a false sense of security to those who use it. Some works on this subject are little more than compilations of advertising for home-security products, some good and some not, but most of which may already be obsolescent or valueless, as their vulnerabilities have become well known. Any work that focuses on the physical aspects of home security alone and ignores the burglar's mind and methods is incomplete. This book was written with all of the above in mind, and aims at educating the reader so that a rational and effective approach to home security can be undertaken. May you enjoy it and use it well!

Mick Davis

CONTENTS

1

YOUR ENEMY AND YOUR VULNERABILITY

Protecting yourself and your property from burglary requires the physical means of doing so and the knowledge to employ it. Knowledge is not tangible, like a door lock, but it is the more important of the two. The lock has no value without your knowing how to use it, but with the proper knowledge you can devise many locks.

Unfortunately, many and possibly most of you reading this book have already been victimized. Most homes and apartments in which no well thought out efforts have been made to improve security are sitting ducks for even an amateur burglar. In spite of all the publicized crime stories, they remain merely stories in people's minds, misfortune that befalls others, but is unlikely to happen to them. They may diligently lock their doors and windows before leaving home and be fully confident that this will keep out intruders; less confident people may consider getting better locks and an alarm system, but procrastinate indefinitely. Some people will go to the trouble and expense of installing good locks and an alarm system but leave them unused because they are inconvenient, or because they are coming home in "only a few minutes." A few minutes is all a burglar needs. No amount of regret can undo the victim's loss and anguish once it is too late. Your enemy is not the criminal alone; it is also apathy and a false sense of security.

PARANOIA AND VULNERABILITY

The United States Justice Department's crime statistics for the year 1982 show both wealthy and poor households to be more likely than middle income ones to suffer theft or burglary. Burglary here includes breaking and entering, where theft is simply the taking of property. Homes in urban areas were the most often victimized, those in rural areas the least. Twenty percent of all American homes were victimized by theft in 1982 and another 7 percent by burglary, with the difference between theft and burglary often one of mere semantics. For those to whom race makes a difference, there were 67 percent more black households victimized than white ones. The fact that many homes were hit more than once is not taken into account in these figures, and no information is available that correlates the incidents to the presence or lack of security precautions. In view of the prevalence of these crimes, the lack of precautions is indeed foolhardy! Nearly 25 million American households—a full 29 percent—were affected by some form of theft or violent crime in 1982, and these statistics exclude homicide and many unreported incidents. The fear felt by many, particularly city dwellers, is not without reason. Paranoia, however, is just as counterproductive as apathy.

Most fear has some basis in reality—that of crime certainly does—but as you confront and act upon the need for self-protection, your fears diminish along with your vulnerability.

AMATEUR AND PROFESSIONAL BURGLARY

The amateur burglar is usually a truly malicious individual. If not, his crimes are infrequent and may be due to overwhelming financial problems. This type of burglar may select his targets with some compassion, preferring the insured business or wealthy home to the ordinary residence. He may be the adventurous sort glorified in fiction; his thefts are not petty, though, and you have every right in the world to be protected from him. He is usually not highly skilled, and reasonable security precautions will force him to strike elsewhere.

The truly antisocial amateur burglar is a dangerous, unpredictable criminal. Because of his lack of expertise he is more likely to be caught in the act than a professional, and may act unpredictably, perhaps violently. Handling an unexpected confrontation with a burglar in your home is discussed in Chapter 4.

The semiprofessional drug-addict burglar is often desperate for the money needed for his expensive habit. His craving compels him to commit several crimes a day, and although experience gives him some skill, he does not have the patience and caution that a true professional uses to avoid being caught in the act. Desperate and possibly irrational because of withdrawal symptoms, this type can easily become violent. Making heroin legally and cheaply available to such addicts by medical prescription would drastically reduce such crimes. One wonders why this is not done!

Unlike most amateurs, professional burglars are not opportunists but cold, calculating criminals who are highly skilled and plan their work well. While amateurs are usually young males, professionals may work with women or children. They often appear to be well-dressed businessmen or uniformed workmen, and are clever in avoiding suspicion even when working brazenly. They have been known to empty the contents of a home into a moving van and escape in broad daylight. Professional thieves can quietly enter your home as you sleep, leaving you to awaken to a loss of several thousand dollars. They possess such finesse that you may not realize you have been burglarized until some valuable is discovered missing days or weeks later. Getting by a doorman or through an ordinary locked door is child's play for them. They may work alone, as a pair, or even in a group. They may carry weapons, but are very careful to avoid confrontation. They have their game down to a science, and if you have something they want badly enough, they are likely to get it in spite of the usual or even better-than-average security precautions. If there weren't easier pickings elsewhere, some of the pros could probably hit Fort Knox successfully. Fortunately, there are relatively few true professionals, but some amateurs imitate their methods.

If the information given in this book is properly utilized you can greatly reduce your vulnerability—even to the professional. It is virtually impossible to render any structure 100 percent intruder-proof, but if you take greater precautions than others in your locality, criminals will almost certainly try elsewhere.

THE BURGLAR'S VIEWPOINT

A burglar is usually looking for easy money. He wants to take as much as he can with as little risk as possible, and therefore wants to work quickly, quietly, and unseen. As he is usually antisocial, he is insensitive to or perversely enjoys the loss, fear, and grief he causes. This is his job and it pays well; his attitude is, "I don't give a damn who disapproves, and if they get in my way they'd better watch out." He figures he will not get arrested, and can get off easy if he does.

The burglar's belief that he will not get caught is understandable. Statistics for a major city in 1983 revealed that only 6 percent of all burglaries led to an arrest, far fewer to a conviction, and still fewer to imprisonment. With insufficient prison space for violent offenders, burglars need not fear doing time; arrest is merely a minor occupational hazard. The U.S. Department of Justice-sponsored National Crime Surveys have revealed that about half of all household burglaries are not reported, which indicates low public confidence in the ability of the police to solve such crimes. In a big city like New York, police will often just take your report over the phone and issue you a case number without personally investigating, unless you are wealthy or politically influential. Just as there are economic and social inequities that contribute to criminality, there are inequities in the services given crime victims. One must look out for oneself.

Our recent history has shown an increase, not only in violent and property crimes, but in "white collar" crime as well, with the distinct probability that the prevalence of the latter contributes to the former. No matter how poorly educated the street punk is, he sees and hears the news reports of criminal activity in high levels of government and business and knows it is only the tip of the iceberg. He loads his gun, grabs his tools, and sardonically sets out to get his piece of the pie. To lecture him on the law would be to beg a violent rebuff. We will get greater integrity among the general public when we have it among our so-called leaders.

The ordinary burglar's motives and methods are the clue to his defeat. If there is not much to steal—that he is aware of—he is less likely to try. Given two opportunities with equal profit potential he will choose the one with lesser risk. Physical barriers are a deterrent and he is likely to give up, at least for the time being, if a job is taking more time than expected. Light from tamperproof sources may deter him if it makes him visible to others, as will the noise of an alarm if it can be heard by neighbors or passers-by. One of the best deterrents is the reputation of a neighborhood or building for being militantly anticrime and patrolled at all times by armed guards, or at least under surveillance by "blockwatchers." Vigilantism has great potential for injustice and is not advised, but given the injustices suffered by crime victims, it is understandable and possibly the lesser of evils where the crime problem is extreme and police service inadequate. Civilian and private patrol services should be undertaken with the knowledge and guidance of the local police, and great care must be taken to exclude unqualified individuals from patrols. Chapter 8 covers citizen patrols in depth.

Guardian Angels founder Curtis Sliwa has said that people have to organize themselves and develop their own protection programs. Some politicians and police oppose citizen patrol groups, but unlike the public they inadequately protect, they carry guns and have bodyguards paid for by our taxes. Long live Curtis Sliwa, Lisa Sliwa, and each and every member of this brave group of volunteers! Curtis's book *Street Smart* covers a great deal more than burglary prevention and is highly recommended.

SECURITY AS APPLIED KNOWLEDGE

Becoming active and informed in any area of uncertainty will help put your mind at ease. So it is with security. The more knowledge you acquire and apply, the safer you, your family, and your possessions will be. Participate in community crime prevention activities and do not hesitate to assist others in this cause. Be alert for possible criminal activity in your area and report anything suspicious to the police and local patrol groups. This can be done anonymously at no risk to yourself. Together we will stand and make this a better world.

PRIVACY AND A LOW PROFILE

Unfortunately, the more successful and affluent one appears to be, the more likely one is to attract all sorts of predators. Ostentatious displays of wealth or luxury are like waving a matador's cape before the bull. You aren't affluent? You may not think so, but the video game system visible through your window may be seen as wealth by a burglar. All those expensive toys your kids leave in plain view in the yard are not exactly necessities, and certain people might wonder what is inside your home if such things are outside. Then there are people like Klondike Pete.

Klondike Pete was a lean, tough old man with a rigidly wrinkled face that must have been battered by many a blizzard. Not quite a hermit, he kept pretty much to himself in a house two miles past the town border. The house was a perfect setting for him—old, cold, and stark. What set it apart from other such houses in the area were a well-maintained fence within which a seemingly vicious Doberman ran freely and all sorts of inviting signs like "Beware of Dog," "No Trespassers," "Premises Protected by Electronic Alarm System" and even "Trespassers Will be Shot, Survivors Will be Prosecuted."

Town gossips, having little else to do, fantasized all sorts of explanations for the nonconformities of those about them, including each other, but of course that was never discussed in the presence of the subject. Pete was said to have been a rugged prospector who made good years after prospecting had lost its appeal. Antisocial types have big ears for gossip, and of course one such with larcenous ways put two and two together and came up with five, as they usually do. That old goat must have a fortune hidden away in that shack! Why else would he be so paranoid? One day, Klondike Pete returned home from shopping to find his gate torn off its hinges, his dog shot dead in the yard, and his home ransacked. Even the floorboards were torn up in a fruitless search for treasure. The only thing missing—probably the only thing worth taking—was his rifle. For the first time since his wife passed away eighteen years before, the old man cried like a baby.

The law generally supports the opinion that one's home is his castle. It is nobody's business but yours how you live in it or what you have in it. It might be necessary to list possessions for insurance purposes, but you have every right to expect the information to remain confidential, and you may have grounds for legal action against anyone who divulges such data. The problem is that our privacy is violated most often as a result of our own carelessness. Unknowingly, we cooperate with our enemies.

Keeping a low profile means not being conspicuous. By all means enjoy what luxuries and comforts you can afford, but don't show them off, boast of them, or let them be easily seen through a window. The millionaire content to live in a comfortable but modest-looking home and drive a good but not flashy car is less likely to be hit by thieves than the playboy that lives extravagantly beyond his means. To flaunt it is to risk it.

As a display of valuable property attracts professional burglars, so do very visible security measures, such as large warning signs and conspicuous window bars. Good protection is done with discretion.

The indispensable telephone is one means of betraying ourselves to our enemies, and they know this and use it well. If there are few people with the same last name in your local phone directory, you should get an unlisted number. If there are many though, you can save the monthly charge for an unlisted number by requesting that your first name and address be omitted from the regular directory and that your listing be removed entirely from the "reverse" (address) directory. Your listing will then be something like R. JONES 123-4567 and will only appear in the standard phone book. If there are a dozen people named Jones listed, you are lost in the crowd, especially if some of them have omitted their addresses as well. Get confirmation of your request from the phone company in writing, and check that it has been complied with when new directories are issued.

* If a telephone caller ever asks your name, address, or what number he or she has reached, hang up without answering, no matter

what excuse they offer for wanting to know. Don't converse with them, just hang up. Make sure that every member of your household knows how to handle this and all other possible telephone security situations.

* If you use a telephone-answering machine, be careful not to let it reveal when you are not home. Just state in your message that you will return the call "shortly," and do not say that you are out. If people call frequently but leave no message, discontinue using the device.

* Never allow yourself to be tricked into revealing to a caller that you live alone or are alone in your home at the time. If you are a woman and a caller asks for your husband, never say that you are single or that he is not home; ask the caller for his name and number and say that he will call back in a few minutes. Children should be taught to handle calls the same way.

* The telephone in your bedroom or security refuge room (see Chapter 7) can be used to call for help in the event of intrusion. It should not be merely an extension, but a second telephone with an unlisted number so that it is not affected if the other phone is tampered with.

* If you ever place an ad requesting that people respond by calling you, never include anything such as "after 6 PM only." Use an answering machine or, better yet, have them reply by mail, preferably to a post office box or mail drop.

* Unless you use an answering machine, you should leave the phone *off the hook* when you are out. Callers will get a busy signal, and if they have it checked will be told there is "trouble on the line." Also, no one outside your door or window will hear your phone ringing without its being answered. Never have your phone temporarily disconnected, as this tells callers you may be away for an extended period of time. When the receiver is off the hook for several minutes, a loud beeping signal is sometimes heard; this is to remind you to hang it up. It can be silenced by putting the receiver under a pillow.

* Many homes have been "cased" by phone. Never leave your home as a result of a call from a stranger; you should immediately report such attempts to the police. Callers may pretend to be conducting a survey or selling insurance or a security service. They are only trying to get information from you. Hang up. Should such callers persist, notify the police and the phone company.

More information on telephone security is given in Chapters 5 and 9.

No more information than is absolutely necessary should be divulged on your mailbox, door, or apartment registry by the bell intercom or mailboxes. Your first initial and last name are sufficient. A great deal can be learned about you by merely observing your mail and garbage, so make sure that your mailbox has a good lock, and be sure to burn or shred any private papers and envelopes before discarding them. When you take delivery of a new TV, stereo, or other large valuable, be sure to remove your name and address from the cartons before disposing of them, and never leave them in front of your home in plain view. Never openly communicate any information indicating when you or someone else will not be home, by postcard, a note on the door, or in a loud voice. If you leave notes on your door or by your mailbox indicating that you are out, write "Burglars Welcome" on the next one, because that is what they say.

It is hard to completely conceal the fact that no one is home, but you need not make it obvious. Carefully observe the outside appearance of your home or apartment when it is occupied and try to make it look the same when it is not. If your blinds are normally open, leave them so. If they are normally open during the day and closed at night, leave them open halfway if you will be out for the full day and night. If you will be out for the evening only, leave them closed if they are usually closed at that time. If your bedroom window can be seen into from the street, a fire escape, or other point, leave the bed unmade and leave other signs that the room is in use.

Lights and a radio left on all the time do not fool anyone; they are a giveaway that no one is home. Interior lights, a radio, and a TV controlled by timers to operate in a realistic on-off pattern, however, are a good deterrent. In hot weather you can use an air conditioner on a timer, set to *low* or *fan* for added realism. If your home is normally dark inside from midnight to 6 AM it should be so, except perhaps for a bathroom light that comes on once or twice for a short time. Gadgets that turn an interior light on for a few minutes whenever they sense a noise should not be used. Any thief with half a brain will recognize their action and realize that no one is home.

The burglar may strike anytime, at any hour of the day, on any day of the year. However, burglars take advantage of major holidays when people are often away from home. If you are out at such times, a trustworthy babysitter or family member should be in the home. At the very least, leave the stereo playing and make it look like the party is in your home, not elsewhere.

Exterior lighting should always be on during dark hours whether you are home or not. Photoelectric controls should be used to do this automatically.

If your home is normally unoccupied during the day, have a trusted neighbor pick up any leaflets or deliveries left at your door, and if you must leave your garbage out for collection, leave it in plastic bags so that no empty cans will remain in front of your home until you return.

Garages should always be kept securely locked. If the garage has windows they should be replaced with sheet metal or strong, solid panels, or at least painted over so that the presence or absence of a car cannot be observed. Seal any cracks under the door or other openings that may provide a means of looking in.

Keeping unpredictable hours, or at least avoiding an easily recognized routine of entering and leaving home, is a good deterrent to criminals. This topic will be discussed in detail in Chapter 9 on carefree vacations; what is given above applies to both ordinary and long term absences.

The dangers in admitting strangers or casual acquaintances to your home is that they may be casing it for a future break-in or they may attack or rob you then and there. The fact that most people are social and trustworthy does not mean that you should allow yourself to be caught off guard because you don't want to seem unfriendly or impolite. The most vicious criminals may seem very kind and harmless until it is too late. You must never let it get that far. Here are some important rules:

* If at all possible, avoid hiring domestic help. If you do hire someone, it is best to do so through a reputable agency, and rather than take the agency's word, do your own background check on any applicant before hiring. Get three or four references (one or two could be covered by the applicant's friends) and check them all. If possible, have consumer credit and police record checks made as well. If you have the slightest doubt, do not hire this person.

* Do as much of your own home repairs as possible, for economy as well as security.

* You should have a wide-angle viewing peephole in your door, but make sure it is closed when not in use. Chain locks are easily broken, so never depend on them to keep someone out once the door is opened. Keep the door locked until you are sure the visitor is OK and that no one is hiding alongside the door.

* Even if you are expecting a repairman, salesman, meter reader, delivery person, etc., insist on seeing identification (have it slipped under the door) and check it carefully before opening the door.

Credentials can be faked. If possible, become familiar with the real credentials that various services use so you can spot phonies. Check the description, photo, and expiration date to make certain the person at your door is the proper credential holder. If suspicious, ask for a personal ID to compare information and signatures. Ask the person questions based on the credentials, such as birth date or social security number. You can call the company he or she claims to be with, but if you have never heard of it, don't bother—it may be a front. Never be fooled by appearances, as a criminal can put on a very convincing disguise, including a police or fireman's uniform.

* A repairman, delivery person, or anyone else with a legitimate reason to enter your home should never be left unobserved for an appreciable period of time. When someone is expected, make sure that no valuables that can be easily taken are left visible, and do not inadvertently reveal where money is kept. Have enough in your pocket to make any necessary payments.

* Never put a "For Sale" sign outside your home, as it gives strangers a good excuse to enter. Deal only through a reputable real estate agent.

* If someone comes to your door to make an emergency phone call, *never* open the door, but offer to make the call for them. Another ruse potential robbers use is to come to your door and tell you that something has happened to a family member. Don't fall for it; call the police to check the information, and if it is false, report the attempted intrusion.

* Never make a public announcement of a wedding, death, vacation, business trip, or anything else indicating that your home will be unoccupied at some future time. Burglars eagerly search for such information.

* An attractive person can be as antisocial as anyone else. Before taking a date home with you, check the person for any negative traits. Don't be blinded by love; someone who turns you on can turn on you. Some rapes, robberies, and worse have resulted from a lack of proper caution.

IF THE WOLF IS AT YOUR DOOR

If you believe that there might be a criminal at your door or loitering outside your house or apartment, these actions must be taken. Never give anyone the benefit of the doubt; it is better to be safe than sorry. If you let a stranger in your home you might find yourself dead wrong.

1. If you have a gun, get it, load it, and make sure the safety is off, ready to fire. Order all family members who are unable to fight to go to a safer location. If you are unarmed, or unwilling to fight but you have a room equipped as a security refuge (see Chapter 7) go to it and lock the door immediately, then proceed to call for help.

2. You can be sure that anyone who would brazenly break into your home in your presence is armed and dangerous. The information in Chapter 4 regarding the use of firearms for home defense (Forewarned Is Forearmed) applies here. The utmost caution is called for under these circumstances. You should become familiar with the state and local laws governing your right to use force in such a situation.

If your doors and windows are strong and well secured, and you do not allow yourself to be conned into opening the door, you will have preparation time and a strategic advantage. *Never* leave the premises to confront a prowler; even if you are armed, this endangers you, your home, and your family. The ensconced defender has the advantage.

3. Double-lock all your doors. Wolves can travel in packs, so be aware that there may be one or more accomplices.

4. Turn on your alarm system if you have one so it will sound if a breakin occurs.

5. If the person at your door has tried to bluff his way in with phony credentials, do not return them. The police will want this as evidence.

6. Call the police and tell them someone is attempting to enter your house or apartment. Make sure to give your full address, apartment or room number, and if possible describe the suspect(s). The police usually record all emergency calls, and it may pay not to hang up, but to allow the recording of subsequent sounds to continue. This could provide valuable evidence in your favor should it become necessary to shoot an intruder in self-defense.

7. If your phone doesn't work, or if your electricity has been cut off, assume foul play. Sound your alarm with the panic button (see Chapter 5) or by opening a window that is inaccessible to the intruder. Summon help in any way you can; scream from the window if you can do so safely.

8. Do not open the door until the police arrive. Put your gun away, with the safety *on,* before opening the door.

As you can see, a gun and an alarm system can be lifesavers. These are discussed in greater detail in Chapters 4 and 5.

PLAYING COPS AND ROBBERS

All members of your household should be impressed with the importance of privacy and caution, and trained in their exercise. If a child can understand why and how to avoid strangers outdoors, he or she can do so indoors as well. A good security game for kids and adults alike is "cops and robbers." Get the family together and take turns finding ways that you would try to get into the house if you were an intruder. While doing this, walk through and outside the house or apartment so that it can be observed carefully. Some funny ideas will come up; list them all. For each idea, take suggestions on preventive measures. Be sure to remedy all security weaknesses that you find before someone else finds them and plays a more serious game. Children have wonderful imaginations, and from the mouths of babes may well come gems.

2

DOORS, LOCKS, AND KEYS

STOP FOOLING YOURSELF

A good deal of our vulnerability comes through self deception. We run in the dark, refusing to believe in cliffs until we fall off one. For instance, look at a front door. Assume it is of solid hardwood construction with both a lock in the knob (bad) and a separate deadbolt lock with a high-security cylinder (good). It fits tightly in its sturdy frame, the hinges are on the inside, and a steel guard plate protects the high-security lock. To prevent prying, there is a strip of angle-iron bolted along the door edge so that no tool can be placed between the door and the frame. Our windows are protected by heavy gates as well, so we complacently go about our business thinking we couldn't be safer if we were the only people in the world. Buy an alarm system? It costs too much, and besides, it would take a tank to get through that door.

A tank? No, just a drill and a hole cutter, plus a few minutes' work. A hole is cut through the wood, a hand reaches in, opens both locks from the inside, and we have learned a costly lesson.

This is not that likely a scenario, as there are probably less well-secured residences nearby that a burglar would choose before going to all this trouble and risk. Such a breakin is possible however, and shows how easily we can deceive ourselves with a false sense of security.

YOUR "CASTLE" AS A PHYSICAL ENTITY

In the recent past, security has not been a major consideration—often almost no consideration at all—in house and apartment design.

This leaves most of us with a sizable but not impossible task. The existing large openings, doors and windows, are the means of entrance in the great majority of housebreakings. They are also the places where efforts at improving security are usually concentrated. Even the simplest of houses has six sides though, including floor and roof, and many small openings exist that are not intended for use by a person but are usable by one. It is of little value to heavily reinforce doors and windows when the outer walls are so flimsy that a few sledgehammer blows can break through them. Some roofs can't even keep out the rain. We'll get back to these areas, but for now let us concentrate on the most often used means of illegal entry.

If you have good tools and some experience in home repair, many of the suggestions made here can be implemented without the cost of hired labor. If you have the slightest reservations about any given task, you should hire an expert carpenter, locksmith, or home contractor. The money you save by doing a poor job yourself will be lost—with interest—if a thief breaks in. There are plenty of incompetents and some crooked people in these businesses, so never choose a tradesman based on convenience or low price alone. Look for several years of experience, a license, membership in a trade association, satisfied customers, and a good reputation with the Better Business Bureau, consumer protection agencies, and police. Check carefully with these agencies and references before hiring help. It *is* worth the trouble.

If you intend to install your own lock, make sure that it comes with clearly written instructions along with any necessary template, and a list of the required tools. Never begin installation until all materials are at hand and you are certain you can finish correctly once you begin.

DOORS AND FRAMES FIRST

Would you secure a lunch bag with a padlock? Of course not, the bag could be torn open around it. You would be surprised how many people spend over $100 on high-quality locks and hardware for a flimsy door, or a door in a weak frame. Don't worry about your locks until you have a door worth locking.

A door frame consists of two vertical members (jambs) with a top crosspiece, and the best frames are of one-piece steel construction. The door hinges fasten to one of the vertical members, and in the opposite member is the mortise, a cutout space into which the latch and lock bolt of the door project. This is covered by the strike plate, a metal plate with rectangular holes for the latch and bolt. Doors in homes should open inward so that the hinges are not accessible from the outside. The frame should overlap both side edges and the top of the door by at least one-half inch on the outside surface. Door sizes are typically 2′4″ to 3′ wide, 6′4″ to 6′8″ high (basketball players hate this) and 1-1/4″ to 2-1/4″ thick. It is crucial that a door fit its frame tightly, with less than a 1/8″ gap on the lock side. The hinges should be of such strength and so securely anchored that a 100-lb. weight hung on the knob produces no noticeable change in the fit or functioning of the door. If it does, the hinges and/or frame are too weak.

There should be very little play of the door in its frame when it is closed and locked. This means that pushing and pulling the knob should result in very little, if any, motion or sound indicating looseness. Excessive play indicates problems with the frame, mortise, mortise hardware, or locks. As far as overall strength goes, a door should be able to withstand indefinitely the repeated blows of a 200-lb. man, using his full body weight thrown against the door. A few such blows are an excellent test; it is better to discover and repair vulnerability if you break the door yourself, than to pay the much higher cost of a criminal breaking it in for you. New construction work should always be tested in this way, and the contractor held responsible for any weaknesses that are revealed.

Door frames are commonly made with 3/4″-thick wood and these are often weak, particularly around the area of the mortise and strike plate. These frames often have considerable space between them and the rough (building structure) frame into which they mount, which makes them vulnerable to prying. Wooden frames also deteriorate with age. Rotted, warped, and softwood frames should be replaced; solidly built hardwood frames, if sufficiently reinforced, provide good security. Hollow aluminum frames are no better than wooden ones, unless they are unusually thick and reinforced with steel plates around the mortise and lock strike areas. Bar, multipoint bar, and buttress ("police") locking devices (discussed later) can provide some reinforcement, but it is best to repair or replace weak frames and doors. The resistance to forcing of wooden frames can be increased by filling the space between the frame and the rough frame solidly with metal or hardwood in the areas behind and around the mortise, and behind the hinges (see Figure 2-1). When this is done, sheet steel should be cut, bent, and installed to overlap and strengthen the wood on the frame edge in front of the mortise, on the inside of the premises (see Figure 2-2). A sheet metal shop can make this

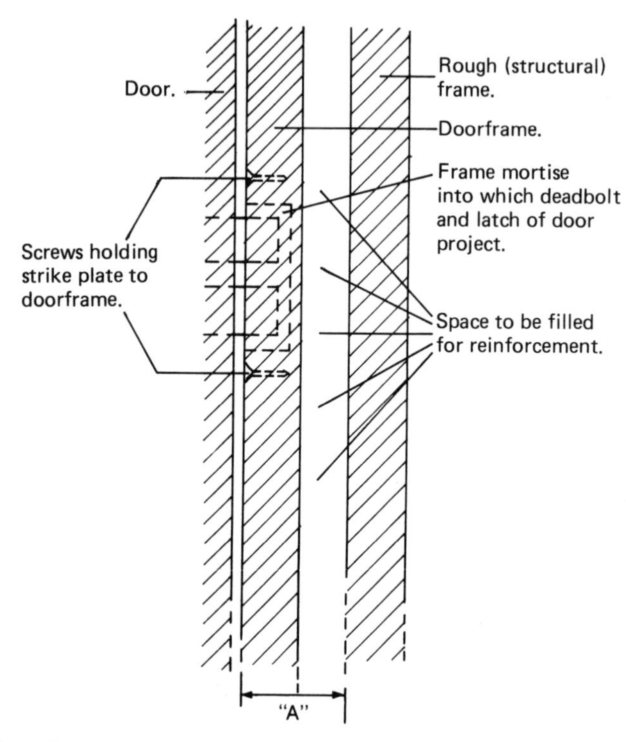

Figure 2-1. Wooden doorframe with front casing (trim) removed for access.

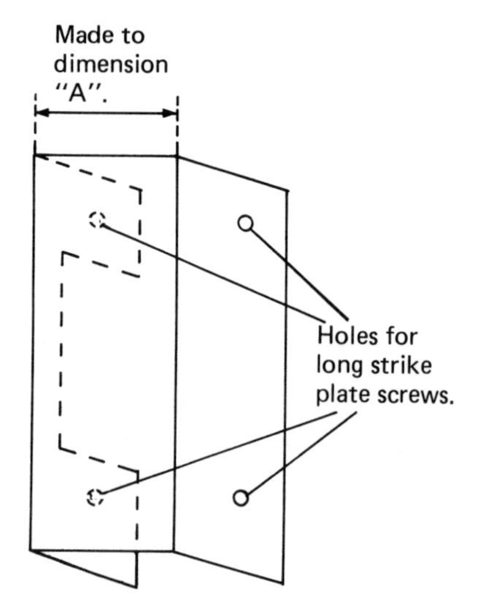

Figure 2-2. Sheet metal reinforcing plate.

reinforcing plate for you from an exact drawing. The strike plate is fastened by long screws that penetrate the newly installed reinforcing materials for greater strength. The frame mortise can also be reinforced with a metal strike box that fits into the mortise and is fastened with long heavy screws (see Figure 2-3).

The best door frames are steel, 12-gauge or thicker (as with electrical wire, the lower the number, the thicker the metal), well-anchored in cement or to the rough frame. Most modern apartment buildings have these.

The cylindrical hinge mechanisms found on the inside of many good doors and frames contain springs that cause the door to slam fully closed when released. If it does not, the springs need adjustment. If the hinges have no springs, an auxiliary door-closing device should be installed.

If a door opens outward, the hinges are probably accessible from the outside; this is a serious vulnerability. Solid, tamper-resistant hinges must be used, and in addition the hinges or the door itself should be "pinned." This means installing steel pins (you can use screws with the heads cut off after insertion) in the door edge or door side of the hinges that fit into corresponding holes in the frame. If a burglar succeeds in removing the hinge axle pins, these door pins will still prevent the door's removal. The best way to use these pins is in place of one of the several fastening screws on each hinge. They should project about one inch, or as much as is practical without interfering with the door's operation.

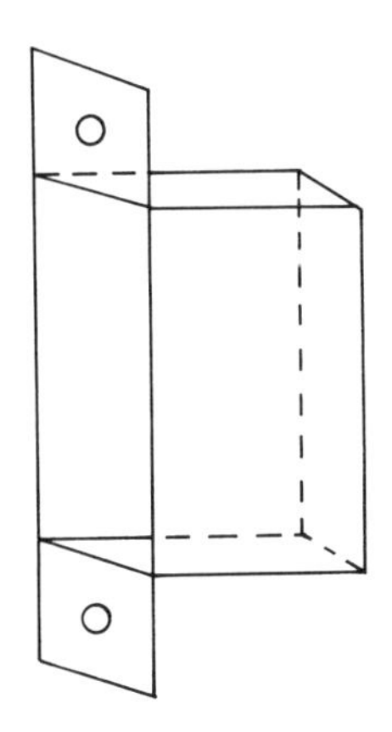

Figure 2-3. Strike box fits into frame mortise, adding resistance to force.

Flimsy screen and "storm" doors offer no protection. A child of eight with a scissors can get through many of these doors in seconds; all they can keep out is flies. Never make the mistake of counting on a screen or storm door for protection from anything bigger than an insect.

Double doors are difficult to secure well unless there is a solid post between them so that each is, in effect, in a separate frame (see Figure 2-4). If not, some protection can be provided by flush bolt locks at both the top and bottom of one of the doors. These are housed in the door itself and should project at least one inch into the frame above and the floor below the door. These are hand operated from inside, so the other of the two doors must be provided with a good interlocking deadbolt rimlock (discussed later) that is key-operated from the outside. This locks it to the other door, that is secured with the flush bolt locks. Rather than flush bolts, square bolt locks that mount to the inside surface of the door are easier to install. They must be large, sturdy, and mounted with good hardware. Some multipoint and bar locks (discussed later) may be used with double doors.

French doors are similar to double doors but are more flimsy. I don't know what the burglary statistics are in France, but if they use these doors the thieves must be rich. Replace them.

Dutch doors with separately hinged top and bottom sections are usually insecure. They are often of flimsy construction and require twice as many locks as a one-piece door for the same degree of protection. Replace them.

When is a door not a door? When it is ajar? No, when it is a hollow core wooden door or one with thin wooden panels or windows. A few good kicks can smash a hole big enough for someone to crawl through or reach through to open the lock. Such doors may be satisfactory for privacy between the rooms of a home, but they are not able to keep out intruders. It hardly pays to try to reinforce such weak doors; replace them. If you are forced to make do with such a door, cover the entire surface of it with sheet steel, 12-gauge if on the outside only, and 16-gauge if on both sides. For fastening, use round head bolts with the heads outside and the nuts on the inside of the door. These bolts have no screw slots and are almost impossible to remove from the outside. Large one-way screws (discussed later) may also be used, and may be best where both sides of the door are covered. Use many of them.

Doors made of solid softwood are marginally better than hollow ones but are quite vulnerable to force as the wood can be easily split.

To test for hardness, try pressing a thumbtack into the door edge; with good hardwood it will be practically impossible to do. Hollow aluminum doors are not much better than hollow wooden ones. Softwood and aluminum doors are not recommended, but if used, they should be reinforced by fully covering the outside with 12-gauge sheet steel.

Windows or panels near a door frame give possible access to the lock; they should be covered with securely fastened sheet metal or grilles on the inside, or at least reglazed with laminated glass or thick polycarbonate plastic. Mail slots, doggie doors, and any opening other than a small, securely installed wide-angle viewer should be eliminated, as they render even the best door vulnerable. Transoms should be permanently covered over on the inside with sheet steel or bars securely bolted to the structure.

Figure 2-4. Double doors. The door on the left is held with square-bolt locks inside at top and bottom. The right-hand door locks to the left one with a mortise deadbolt, but the exposed bolt between the doors and the cylinder are vulnerable; protective plates should be used here. The alarm (tape on glass) adds security, but a pro could probably bypass it. The chain link fence seen reflected in the glass has sharp wire points at the top to discourage climbing, but anyone really determined to do so can get over it.

Doors of solid hardwood or particle board are satisfactory if properly reinforced with sheet steel as mentioned above on at least one side. If the sheet does not cover the entire surface of the door so that the frame overlaps it when closed, it should be on the inside. However, an outside, full-sized sheet is still best. Because of the space cut out in the door for the mortise lock and knob or handle mechanism, this area is a weak point and sheet steel reinforcement on both sides here is recommended. This reinforcement work should be done even if strength-adding devices such as buttress or multipoint locks are used. Security should never be sacrificed for the sake of appearance; even sheet steel can be attractive, properly painted and finished.

Careful planning should precede the reinforcement of a door of any type with sheet steel. Would an outside full-sized sheet interfere with the door's operation because of its thickness (12-gauge steel is 1/8 inch thick, 16-gauge 1/16 inch thick)? Are the hinges strong enough to bear the weight, roughly 100 lbs. total, for the sheets recommended above? Could you do the work yourself, or would the cost of materials plus labor make it more practical to replace the entire door with a steel one? Should you decide to proceed, make an exact drawing. Indicate the location, shape, and dimensions of any holes or cutouts in the sheets that are necessary to accommodate the peephole viewer, knob or handle, and lock cylinders. Do not let the sheet extend over the lock cylinder, leaving only a small hole for the keyway, unless the cylinder can be replaced from the other side of the door. If the sheet cannot protect it, use a separate guard plate. When you are satisfied with the accuracy of your drawing, take it to a sheet metal shop (see your Yellow Pages) and get a price for the job. It often pays to shop around. If acceptable, proceed. Use only round head bolts or one-way screws on the outside for fastening, and NEVER skimp on the hardware.

The best doors are steel, 16 gauge minimum thickness, in good steel frames. With high-security locks properly installed, these are a formidable obstacle to any potential intruder.

Doors often come with an ordinary latch and deadbolt lockset or even a flimsy key-in-knob lock. You are better off getting a door without this and putting in good high-security locks of your choice (see Figure 2-5).

Solid steel roll-up doors, when properly anchored and used with high-quality locks, provide excellent protection but are more practical for storefronts than for homes, and when equipped with motorized

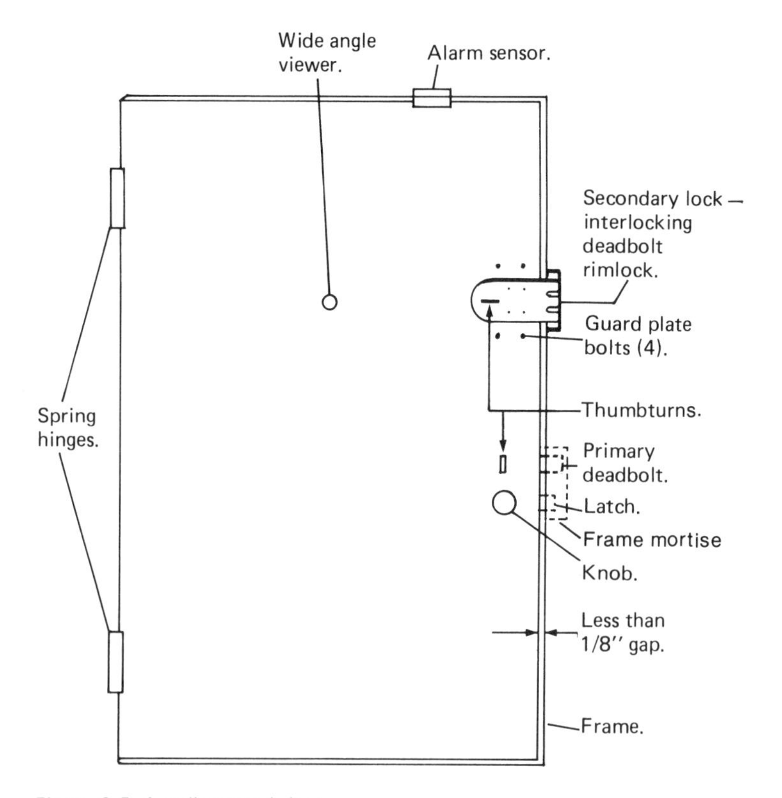

Figure 2-5. A well-secured door.

openers, make excellent garage doors. Open mesh roll-up grilles and folding gates are inferior to roll-up doors of solid, flat steel construction that cannot be gripped easily by a tool or chain to force them open. If such a door has an outside lifting handle, it should be constructed so that excess force will break it off before the door itself can be forced up or out. When provided with exact dimensions, a dealer can give you an estimate of the cost of such a door, including installation. Check the Yellow Pages under "DOORS" and "GATES." Remember that no door is any stronger than its locks. If exterior locks must be used, high quality shackleless lock-hasp devices (discussed later) are preferable to ordinary padlocks (see Figure 2-6).

Sliding glass doors as found on apartment terraces and at ground level in private homes are vulnerable in three ways:

1. The latch or other locking mechanism may be defeated by skill or by force. This is often very easy to do.
2. The door may be lifted and removed from the track in which it slides.
3. The glass may be cut or broken.

That these doors are highly vulnerable is well known to the burglar, and even if your terrace is on the fifteenth floor of a thirty-story building, he can get to it. Unfortunately, as such doors are often located in places not easily visible, the criminal can work with little risk of being seen. Without effective preventive measures, these doors are a weak point in your home security. There are many variations in the construction of sliding doors; in some cases there is one fixed glass segment and one sliding one, in others more than one segment can move. The measures taken must be appropriate to the construction, as what would be effective in one case might be ineffective in another. Give careful thought to how your sliding doors operate and how

Figure 2-6. Steel roll-up doors and window shutters are excellent protection but are no better than the padlocks used if they lock from outside. The accordion/scissor gate to the right is more vulnerable.

effective any particular measure will be. Look at them from the burglar's viewpoint, then outwit him.

The latches most commonly found on sliding doors are inadequate. One means of preventing the opening of such a door is to place a thick dowel, pipe, or even broomstick cut to the proper length in the bottom track. This is ineffective, however, if there is any space through which a tool can be inserted from outside to remove it.

A "Charlie bar" locking device can be used to prevent forced opening; this is available at hardware stores and from locksmiths. If possible, this is best installed midway between the floor and ceiling, bracing the center of the door edge against the wall or frame.

Key-operated sliding glass door locks are available that protect against both the forced opening and the lifting of the door from the frame. Professional burglars are familiar with the vulnerabilities of commercial products, however, and a simple but well thought out do-it-yourself lock may be better, as well as cheaper.

There is a certain amount of space, perhaps an inch, between the top of a sliding door and the top of the track above it. To remove the door from the lower track, a thief can lift it upwards into this space. If it cannot be lifted, it cannot be removed. Install several sheet metal screws into the top of the track to prevent lifting. They should protrude far enough so that there is only about 1/8″ clearance between their heads and the top of the door. Use several screws at different points along the track. Precautions against the removal of fixed as well as sliding sections are necessary in some cases.

Another way to secure sliding glass doors is similar to putting a bolt into a hole through the sashes of a double-hung window to prevent their being able to slide open. Drill holes through both the sliding frame and a stationary part of the frame or tracks. Insert tight-fitting steel bolts or pins to prevent both the sliding open and the lifting of the door. These bolts or pins should be at least 1/4″ in diameter and must be located so that they cannot be seen or removed from the outside.

None of the above measures can protect against glass breakage. The doors should be constructed of heavy laminated glass and be equipped with alarm system glass breakage sensors such as foil tape. Alarms are an excellent deterrent; see Chapter 5.

Steel patio gates add security to sliding glass doors, but are costly and should not be relied on alone.

Any door that is never used is a weak point in your home security. Bolt it, bar it, cement it, or best of all, replace it with a brick wall that is as strong as your other walls.

Cellar hatch doors should be constructed of heavy steel, securely anchored in concrete, and barred from the inside. An outside padlock can be used as a backup to the inside lock, but make sure it is a good one; more on padlocks later.

If you have a garage, be aware that some burglars specialize in garages alone. Doors, windows, and all possible points of entry must be as well secured as those of your home. Many garage doors have inadequate locks and thin, easily broken panels or windows. Garage doors that swing open often have exposed outside hinges and weak padlock hasps. When provided with tamper-resistant pinned hinges and a very good padlock and hasps, heavy swing-open doors are more secure than many overhead doors, which are not easy to secure from the outside alone. The best garage doors are the steel roll-up type mentioned previously.

The best garage door locks are fastened on the inside, but this causes the inconvenience of going around through another door to lock and unlock the garage. From the inside, overhead doors may be secured with bar devices or padlocks that go through holes in the tracks to prevent the doors from rising. If you have an overhead door with no inside manual locks on it, you should have an automatic door-opening system operable from inside your car. The best type uses a coded signal to prevent operation by ordinary radio signals or by a burglar with sophisticated equipment. A backup battery power supply is a good feature, as it permits the opener to work in the event of a power cutoff. Make sure to keep the transmitter secure, as you would a key, and never leave it in your car or anywhere it could be stolen. With this, you can open the door, turn on the garage light, make sure no intruder is present, drive in, and immediately shut the door behind you.

Check the Yellow Pages under "Door and Gate Operating Devices" for retailers of garage door openers. See publications such as *Consumer Reports* and *Consumers' Research Magazine* for the latest objective evaluations of available products, features, and prices. These will help you get the most for your money; buy what you need, not what the retailer wants to sell you. The shopping hints for windows and gates in Chapter 3 apply in part to garage doors and openers as well.

If your garage is attached to your home, make sure that the door between them is a very secure one—at least as good as your outside doors. Use high-quality locks that require keys to open from the garage side, and connect the door to your alarm system with a sensor on the home side.

Many key-operated garage door openers, alarm system on-off controls, and other security devices are operated by a keyswitch that makes a momentary contact. For greater security, add a second switching device connected in series with this. Both will have to be operated simultaneously to activate the controlled device; this would require either two keys, or the operation of a cleverly concealed contacting device while one key is used. You should be able to operate this inconspicuously. With the hidden device in use, tampering with the one visible keyswitch cannot have any effect.

Attic doors and hatches are often overlooked, but they are as important to secure as the door between the garage and home, and for the same reasons. If a burglar gets through your roof or skylight, this door may be your last defense. Make sure that it is connected to your alarm system with a sensor on the inside, inaccessible from the attic. Unless your basement has no windows, outside doors, or chutes, this applies to the door between the basement and house as well. It is not essential that basement and attic doors be secured with key locks; a large, sturdy hand-operated sliding bolt lock is satisfactory if it is properly mounted on a good door and frame. A heavy square bolt lock is far superior to the flimsy "barrel bolt" for this purpose. Be sure to use good hardware when installing it (see Figure 2-7).

Figure 2-7. A hand-operated square-bolt lock mounted on a steel fire exit door. As the doorframe is also heavy gauge steel, a small hole into which the bolt projects is an adequate strike.

Take great care that you NEVER store or leave tools, ladders, weapons, or anything else that a criminal can use against you in the garage, an accessible basement, attic, shed, or outdoors. These are best kept in a locked closet, storeroom, or cabinet in your house.

While many landlords are responsive to tenant's needs, some see tenants as nothing more than money sources, to be given as little as possible in return. If you own your house or apartment you have the freedom to replace or reinforce your doors, frames, and other structures as you see fit. If you are a tenant you can run into problems, the least of which is putting back the old door and taking your new one with you when you move. If you intend to stay in your rented residence for several years, it might pay to get the landlord's consent (and get it in writing!) based on the fact that you are improving the property and will leave it so improved when you vacate. If your landlord is intransigent, you should check local laws and building codes to find which he is violating and use this as leverage in dealing with him. Your weak door, rotted door frame, or windows may be something he is legally obligated to repair, but he may take advantage of your not knowing this. Even if you have signed a lease agreeing to do your own repairs, this part of the lease may be illegal and unenforceable. Learn your rights and demand that they be respected. Chapter 8 deals with the security problems of tenants and how to handle uncooperative landlords.

LOCKS, KEYS, AND THEIR PROTECTION

There are far more locks and related hardware made for good appearance than for maximum security. Do not be fooled by decorated, polished brass junk. You have too much to lose.

Various security devices, such as lock cylinder guard plates and angle iron strips that are bolted near the edge of the door to prevent the use of a prying tool are useful and often necessary. They are of little value, however, if they can be removed easily. Nuts and bolts of good quality are as essential as doors and locks of good quality. Only slotless, round headed steel bolts and one-way (nonremovable) screws should be accessible from the outside, and heavy bolts are the most secure. Mounting and fastening methods in which no hardware at all is visible on the outside are even better. One method is to use studs welded to the device to be mounted and held by nuts on the other side of the door. Welding the device itself to a steel door or frame is good if there will never be a need to remove it. The escutcheon plate around the knob and mortise lock and the knob itself should be fastened by invisible means, or at least be very difficult to remove. If two-way security such as that provided by a double-cylinder lock (one requiring a key inside as well as outside) is desired, this applies to the inside fastening hardware as well.

One-way screws and bolts have slots shaped so that a screwdriver can turn them clockwise for insertion but not counterclockwise for removal. They can be removed with a hammer and chisel, but the noise it makes and the time it takes make it unlikely that a burglar will attempt it. Ordinary slotted round head screws and bolts may be made nonremovable by grinding down the heads after insertion to completely remove the slots. These are even harder to take out. Slotless, roundheaded bolts such as those used to mount a guard plate are often held by easily removed nuts on the inside of the door. For two way protection, they can be made nonremovable by destroying the bolt threads after installation with a wrench, chisel, or other tool, or by coating the threads with epoxy glue. Of course it will be very difficult for you to remove the bolts when necessary, but you can make noise and take your time; the intruder usually cannot.

If you must fasten any locking device to a wooden door or frame and cannot use bolts, use screws that are threaded over their entire length and are as long and thick as is practical. When the screw is almost fully in, take it out again, wet it with wood glue, reinsert it, and tighten it all the way. This gives it a more permanent and solid grip.

A worn out or "stripped" screw hole in wood can be repaired, or you can use a larger screw that will fit it tightly. The best way to repair it is to insert glue in the hole and then hammer in a tight fitting wooden peg. Several matchsticks can be used in place of a peg. When the glue is dry, trim off any protruding wood, drill a small guide hole, then insert the screw, wetting it with glue before the final tightening.

Primary locks are those that come with the door and are part of the knob or handle mechanism. They are generally not as secure as secondary locks such as a separately mounted interlocking deadbolt rimlock (see below) and should not be relied on alone. Primary locks usually have a spring latch which closes when the door is slammed. Latches are very vulnerable, and some of the "deadlatching" devices, such as a small bar against the latch, are often ineffective against shims such as a plastic strip. A true deadlatch lock cannot be opened in this way. Fortunately, most primary locks have, in addition to the latch, a separate deadbolt which is key operated from the outside and projects into the door frame through the strike plate. The better primary locks are generally "mortise" locks, as they are housed in a mortise in the door, and operate with a mortise in the frame. They are sometimes referred to as "interconnected locksets."

Key-in-knob locks are junk. Even the best of them cannot compare with a good single cylinder mortise deadbolt. They are inherently weak and should be replaced. To even amateur burglars they say "try me first," and they will, with success.

Another poor bet is the old lever-tumbler lock with the big keyhole you can see through if it is open on both sides. This is easily picked or opened with a "skeleton key."

Here is what to look for in selecting a primary lock:

1. A deadbolt separate from the latch with a minimum length of one inch when extended, constructed of case hardened steel and/or containing a saw-resistant steel pin or ceramic insert.
2. No vulnerable hardware such as ordinary screws exposed on the outside.
3. Protective hardware such as a tapered, free rotating wrench-proof ring to protect the key cylinder from twisting or pulling.
4. Heavy steel construction.

Always keep in mind that although the latch automatically closes, the deadbolt does not. If you do not lock it at all times, it is of no value. Locks with buttons that permit the latch to be opened with the outside knob or thumb latch are best avoided. Going out, even for only a few seconds, with the door "on the latch" is a dangerous practice. If you have this kind of lock, cover the buttons over with strong tape (see Figure 2-8).

Some locks come with cylinders that are the weakest link in the protection they provide. It may pay to select a better cylinder; more on this later.

Even if you have a very good primary lock on your door, a secondary (auxiliary) lock should be used. Cylinder deadbolt locks (deadlocks) which mount in the door require an additional mortise to be cut in the frame for the bolt, a hole through the door for the cylinder, and another hole in the side of the door for the bolt. With a good quality door and frame this is a good deal of work, and doing it will weaken the door, so the type of secondary lock that fastens to the inside surface of the door and frame is advisable. These are sometimes called rimlocks or surface mounting locks. Both the lock that mounts on the door and the strike that goes on the frame must be fastened with strong hardware that cannot be ripped out by brute force.

To be most effective against force the secondary lock should not be located too close to the primary lock; it should be placed about midway between the primary lock and the top of the door. Optimum lock placement is at approximately 40 percent of the door's height for the primary lock, 70 percent for the secondary lock.

Chain locks are not locks, they are a means of deceiving yourself with a false sense of security. If there is anything you *don't* need, it is a chain lock.

Deadlatches, sometimes called slam locks, are among the less secure types of auxiliary lock. These have a spring-loaded, beveled latch which enters a strike mounted to the frame. These lock automatically when the door closes, but have some of the weaknesses of the primary lock latch. The better types are constructed so that the latch cannot be opened with a shim or similar tool. Properly installed on a good door and frame, many deadlatches can provide better security than the typical key-in-knob lockset, but are far from being really good protection. This is unfortunate, as such locks are the only type suitable where automatic locking is required, as with apartment building entrances.

Figure 2-8. A good primary lock in a steel door. Above the extended deadbolt are the two cylinder setscrews, below it is the beveled latch. A piece of tape covers the buttons to prevent leaving the door "on the latch" (openable with the knob from outside).

One of the better types of surface mounting latch locks has three hooks on the strike that is mounted to the door frame; these engage a vertical latching bolt in the lock and shield it from tampering attempts. Unfortunately, the strike hooks are vulnerable to force; if they were constructed of steel this lock, properly installed with a high-security cylinder and protective hardware, could provide truly good protection (see Figure 2-9).

Auxiliary deadbolt locks are superior to most latch locks. These project a heavy, usually rectangular bolt from the lock body into the strike on the door frame. The quality specifications given above for primary lock deadbolts apply to these as well.

The best of the common types of auxiliary locks is the interlocking deadbolt, also known as a vertical bolt or dropbolt lock. This projects a bolt vertically that goes through round openings in the strike. This all-around grip is impenetrable by many methods that succeed with latches and deadbolts (see Figure 2-10).

Figure 2-9. A surface mounting deadlatch with a three hook strike. The bolts through the door hold the cylinder guard plate.

Some surface mounting locks look sturdy but are made with thin, weak casings, or mechanisms that are easily broken. Check *Consumer Reports* and other independent testing agencies' findings carefully before making your selection.

Many otherwise secure secondary locks come with short wood screws or other inferior hardware that can be defeated easily by force. Discard these and use superior ones, as recommended earlier.

Installation instructions are generally supplied with surface mounting auxiliary locks along with a template for properly positioning the holes to be drilled. The general procedure is outlined below; if the instructions with your lock differ from this, follow your instructions unless they are obviously wrong.

1. Locate the secondary lock about halfway between the deadbolt of the primary lock and the top of the door. Tape the template to the door here as instructed, and with an awl or punch, mark the centers of the cylinder hole and the mounting screw holes through it.

Figure 2-10. A double-cylinder interlocking deadbolt. The nuts and bolts above and below it hold the guard plate.

2. Remove the template. With a hole cutter of the proper diameter, cut the large cylinder hole completely through the door.

3 Drill the smaller holes as called for in the instructions; these *do not* go completely through the door. Do not drill the holes in the frame for the strike yet.

4. The key cylinder has a tailpiece of thin, flat metal protruding from the back. This, when turned, operates the lock. This is notched so that it may be broken off shorter, if required, to fit the thickness of your door. Never do this until absolutely certain that it is necessary. A metal retaining plate with holes for mounting screws and a larger central hole for the cylinder tailpiece is used to mount the cylinder. Fasten this to the inside surface of the door over the cylinder hole as directed.

5. A guard plate (discussed later) that is mounted independently to the door offers the best cylinder protection. However, if you wish to use (or the lock comes with) a protective ring or cylinder "rose," place the cylinder through this, then put it into the door from the outside. Two screws that go through the retaining plate enter tapped holes in the cylinder to hold it in place. (Like the cylinder tailpiece, these are usually made so they can be broken off shorter if necessary.) Tighten these firmly. The tailpiece should project through the center of the retaining plate and rotate freely when turned with the key in the cylinder.

6. You are now ready to fasten the body of the lock to the inside of the door. These locks generally have a spring-loaded metal plate or similar device that is intended to prevent a burglar's access to the lock mechanism if he succeeds in removing the cylinder by force. They are often more troublesome to the installer than to the burglar. Follow the instructions that came with the lock. It may be necessary at this point to shorten the tailpiece.

7. With the lock held firmly in position against the inside of the door, use the key from the outside to make sure the tailpiece properly engages the mechanism and the lock operates as it should. If so, insert and tighten the mounting screws. When done, check for smooth operation with both the outside key and the inside key or thumbturn. On some hollow sheet steel doors the overtightening of the mounting screws for the cylinder or lock body can bend the door's surface so much that the tailpiece rubs on the retaining plate as it turns, causing difficult operation. If this happens, loosen the screws slightly.

8. Install the strike on the door frame. Even if the template located the screw holes for this, do not drill them until you are certain they are positioned properly. This is especially critical with an interlocking deadbolt strike where the projecting flanges of the strike

must fit into the lock body, and the vertical bolts of the lock must go exactly through the holes in these flanges.

9. If the door frame is steel or solid hardwood, several large screws may be adequate in fastening the strike. If after drilling only an inch or so into the wood, the drill goes through into an open space, the frame is too weak to securely hold the strike and will have to be reinforced, as discussed earlier and illustrated in Figure 2-1. Occasionally with a wooden frame it is necessary to recess the strike slightly so that the door can close and lock flush against the frame, with no gap on the outside that can admit a prying tool. Carefully use a sharp chisel to do this, removing no more wood than is absolutely necessary.

If you want a maximum security lock that provides physical reinforcement for the door—or you must make do with only one lock, use a bar, multipoint, or buttress unit. These are sometimes called "police locks" (see Figure 2-11). One lock of this type has its cylinder at the center of the door. When this turns, it operates a gear which engages heavy bars above and below it, sliding them into or out of two brackets that are mounted to both sides of the frame. Multipoint

Figure 2-11. A buttress type "police" lock with its bar.

locks often have cylinders in the usual location, but rather than operating a single bolt with one strike, it operates several. These bolts are usually at the top and bottom, as well as the sides of the door. The buttress lock props a steel bar from the door to a receptacle in the floor at a distance in front of it inside the residence, and can only be used with inward opening doors. They are vulnerable to tampering with a tool inserted under the door, so if used, the door must have no such space or else have a protective plate covering any that exists. Bar and multipoint locks are preferable.

Some doors are made with a built-in four-way multipoint lock which projects bolts into the frame at the top, bottom, and both sides. If the door is of solid sheet steel construction and a guard plate protected high security cylinder is used, this is an excellent arrangement.

A double cylinder lock provides additional security, especially if an intruder attempts to penetrate the door to open it from inside. If he succeeds in entering by another means, it can block his escape through the door with heavy loot he cannot carry on a fire escape or get through a window. This lock requires a key to be opened or locked from both inside and out. The key for the inner cylinder should be well hidden and never left in the lock, but its location must be known to all family members should a quick exit be necessary in the event of a fire.

The cylinder is the cylindrically-shaped part of a lock that is operated by a key to open or lock the door. They can be defeated in a number of ways, including picking and removal by force. Fortunately, they are easy to replace with better cylinders that are more resistant to attack.

When to change your lock cylinders:

* Before moving into a new house or apartment, even if it is newly constructed.
* If any evidence of tampering or damage to an existing cylinder is found.
* If present cylinders are old, or of questionable quality.
* If a key has been lost, or is missing.
* If any unauthorized person has had even brief possession of the key.
* If a domestic employee quits, or is fired.

When in doubt, change the cylinder, pronto! This is quite easy to do with a mortise lock. On the edge of the door alongside the cylinder you will find one or two recessed setscrews that enter a groove in the side of the cylinder so that it cannot be easily removed from outside (see Figure 2-12). If these setscrews are not visible you must remove the faceplate on the door edge to get access to them. When the setscrews are loosened sufficiently the cylinder may be removed from outside by turning it counterclockwise. It may be necessary to twist it a bit with a key or screwdriver in the keyslot to get it started. Take it with you so you can buy a replacement of the correct length, but don't leave your home unoccupied without a working lock. The new cylinder must have the same type of end cam; if not, you can usually replace it with the one from the old cylinder. Screw the new cylinder into the lock and tighten the setscrews, making sure that the cylinder is straight (the keyslot is vertical) so that the setscrews enter the groove in the side of the cylinder. Now check with the key for proper operation; it may be necessary to screw the cylinder further in or out for the cam to engage the lock mechanism. If all is well, tighten the setscrews. It is a good idea to put glue or tape over them to prevent their surreptitiously being loosened by someone planning a future break-in. This must be removable for future cylinder replacement, but should prevent quick access to the screws (of course, this is not necessary if they are protected by a faceplate). A good time to lubricate the lock is when the cylinder is out; lock mechanisms and cylinders should be lubricated from time to time with a graphite lubricant or WD-40, but never with regular oil. It's a good idea to keep a few spare cylinders on hand, should a quick change ever be needed.

Figure 2-12. A high-security mortise lock cylinder showing the end cam (held by the two small screws). The grooves in the sides of the cylinder are for the door setscrews. To the right is a beveled guard ring.

Cylinder changing in secondary locks is often just as easy. If the lock is the surface mounting type you must remove the lock from the door for access to the cylinder fastening hardware. If you have trouble, get a good locksmith.

Lock cylinders are constructed with metal parts (pins) that must be properly positioned with the key so that the lock can be operated by turning. Rather than replacing a cylinder, you can have a locksmith rekey it by changing the pins. This is hardly worth the small savings though; if a bad job is done, the lock will be more vulnerable to picking than before. You are far better off investing in a new high-security tamper resistant cylinder.

The selection of locks and cylinders should be made by referring to the most recent objective evaluations, such as those of *Consumer Reports*. Be skeptical of ads, retailer's opinions, and the evaluations of publications or individuals influenced by profit motives. Medeco products are very good. There are others, but let the buyer beware!

The best cylinders require specially cut keys that are not easily duplicated, but this does not include the cylinders that use short tubular keys. The ones of hardened steel construction are the best of this type, but in general this type of cylinder is not recommended.

Cylinders made for the use of a master key, as well as a private key, are highly vulnerable to picking. Replace any such cylinders immediately, and if the landlord or other keyholder objects, make sure you know your rights and insist on their being respected. Nobody has a right to cause you to be vulnerable to crime. If your cylinder is not master keyed but the landlord or superintendent has a key, change it immediately and *do not* give them the new key. Rarely are you legally required to do this. Such keys can easily be stolen or duplicated and used to burglarize you. If you do not want to fight an insistent landlord, and you are not obligated by law to comply, give him keys from old, discarded locks that can enter the keyslot but not operate the cylinders of your present locks. If you fear accidentally locking yourself out, leave an unmarked set of keys with a good, trustworthy friend or relative but NEVER, under any circumstances, hide any key outside the premises.

The only possible legitimate reason for a landlord or superintendent wanting access to your apartment is in case emergency repairs are needed when you are out. In general, especially in newer buildings, this is far less likely to occur than a burglary attempt. Should an emergency occur, the landlord usually has the right to break in. If you wish to or are legally obligated to permit landlord access, give the landlord the phone number (but not the address) of the friend or relative who has your keys. This person should meet them at your apartment, let them in, watch them like a hawk, and securely lock up when they leave. The keys should never leave this person's hands. Remember, even if you have signed a lease which requires you to leave keys with the landlord or superintendent, that part of the lease is unenforceable if the law contradicts it.

There are "keyless" locks operated by magnetic cards and a plethora of similar devices, but many, while cleverly designed and secure in some ways, are vulnerable in others. To be avoided are locks opened by entering a combination of digits or other code; it is too easy for a thief to observe you from a distance to learn how to get in.

To avoid educating the wrong people, the various ways in which a lock can be defeated will not be detailed here. Just be aware that there are ways through even the best locks, so make sure you install protective hardware to thwart such attempts. Steel guard plates mounted with bolts through the door should be used for auxiliary lock cylinders and primary locks which they can fit. They are easy to install; just place the plate on the door over the cylinder so that the keyslot is accessible through the hole, mark the four spots for the bolts, drill the holes through the door, and mount the plate. If such a plate cannot be employed, a free-rotating beveled cylinder guard ring made of hardened steel should be used. The thin brass rings that are supplied with many cylinders are almost useless. The best protection against prying and similar physical attacks is a good door, fitting tightly in a good frame with no free play. Added protection may be had by bolting angle iron strips to the door abutting the frame so that no tool can be forced between them. Such strips should cover the entire length of the door or at least extend for two feet above and below the locks (see Figure 2-13).

All household members, especially children, must be educated in the importance of key security and the necessity of using all locks even when home or going out for only a few minutes. It may be preferable to let a neighbor care for a young child after school until you get home than to let the child have a set of keys. Make no more copies of your keys than are absolutely necessary, and never let them out of your sight, as an impression can be made in seconds from which a good key can be cut. If your key has any code numbers on it, write them down for future reference and file them off the key. A key can be duplicated using that information alone. Keep your keys in your front pants pocket, not in a jacket, coat, or purse from which they may be stolen without your knowing it. Keep your car keys on a separate key chain or ring, and *never* tag your keys with your name, address, license plate, or any other identifying information.

In general, padlocks and hasps (the hardware connecting the lock to the door) are not as secure as good mortise or interior surface mounting locks, and the latter are preferable wherever possible. The thin cased, flimsy combination locks often used for lockers are easily broken and are low security at best (see Figure 2-14). The same is true of the inexpensive locks with flat keys stamped from sheet metal. They are practically useless. If you must use a padlock, look for these features:

* Heavy, corrosion-resistant body of solid steel, or brass reinforced with hardened steel pins.
* Case-hardened steel shackle at least 7/16″ thick, with "heel and toe" locking action (the lock grabs the shackle internally at both ends).
* High-security pick resistant cylinder that does not release the key until the shackle is in the locked position.

Figure 2-13. This door and its secondary lock are protected by a guard plate and an angle iron strip.

Padlocks with a shrouded shackle are expensive but effective. These are made with the body of the lock extending upward around the shackle to protect it from cutting or forcing attempts.

It makes no sense to use a good padlock with an inadequate or improperly installed hasp. A good hasp is made of thick hardened steel and is constructed to cover its mounting hardware when closed with the lock in place. The hinges are pinless and the staple (through which the lock passes) is hardened steel at least as thick as the lock shackle. It should be fastened with bolts through the door, never wood screws or nails, and if on a wooden door, the door should be reinforced on both sides with sheet steel plates where the hasp bolts go through it. Some excellent but expensive lock-hasp devices constructed without an exposed shackle are available; check security equipment suppliers for these. These are the ultimate in padlocks, as the shackle, hasp, and fastening hardware are totally inaccessible when the lock is in place (see Figures 2-15 and 2-16).

There are times when something of value that can be easily stolen must be secured out of doors or in a place readily accessible to others.

Figure 2-14. Poor security padlocks. What good is a hardened steel shackle when the lock's case, cylinder, or locking mechanism can easily be defeated by force?

Figure 2-15. The steel reinforcement plate was welded to the door, the hasp staple to the plate, and the hasp itself to the steel doorframe. Marks at the top of the case of the high security padlock show that an unsuccessful attempt was made to force it open.

Figure 2-16. High security for a cellar hatch door. When the lock is in place its shackle, the hasp, and its fastening hardware are all inaccessible.

Bicycles, motorcycles, boats, carts, trailers, wheelbarrows, and property in a storage room all require special measures to safeguard them against larcenists. Products are on the market designed for specific purposes such as locking a bike to a pole or an outboard motor to a boat. Some of these are fairly effective, others practically worthless. The wise consumer must do some homework by seeking out objective evaluations before making a purchase. The padlock and chain is a commonly used means of security, but is not as safe as a well-secured garage or residence. For the best protection, use both. Remember that it is a waste of good hardware to chain something to a fence, pipe, or other anchor that is weaker than the chain or lock. Motorcycles chained to parking meters have been stolen by removing the head of the meter and slipping the chain up and off the pipe. Chaining to a lamppost or a tree with a very thick trunk is best.

Stranded steel cable is relatively easy to cut, regardless of thickness. For good security, use case hardened welded link steel chain with a minimum 3/8″ link thickness. Even this can be cut with the proper tools, but it takes time, and this discourages attempts. It is best to have the chain and lock positioned as high above the ground as possible, as this makes cutting more difficult.

The story is told of the recently widowed elderly woman moving to the city from her country home to be near her one surviving son. Unaccustomed to urban life, she was deeply shocked when within two weeks of her arrival, she came home to a ransacked apartment. She had little worth taking, but her radio—her sole full-time companion against loneliness—was gone. The villain was particularly adept at opening ordinary locks without damage. This lowlife either had the world's biggest key collection or was an evil reincarnation of Houdini. Her son's replacing the radio was little consolation, as now she was in mortal fear of the thief's returning when she was home. The building superintendent, friendly and always willing to help for a small sum, changed the lock cylinder and installed an auxiliary deadbolt lock. The peace of mind these provided lasted only three weeks, when both locks were expertly opened and the new radio stolen. As she was to soon receive an inheritance and insurance benefits, the material loss was a minor blow compared to the renewed fear for her personal safety. At this point, her son hired a locksmith to install two additional deadbolts so that four separate keys were required to operate all the locks. Two of the four cylinders had to be turned counterclockwise to open and the other two clockwise. The locksmith, affecting the wisdom of Solomon, advised her to lock only one of the two of each type upon leaving, and to leave her radio playing so it would appear that she was home. Any remaining doubts were dispelled when the next burglary attempt succeeded only in locking the two unlocked locks while opening the two that were left locked. Confident of her security, she furnished her apartment comfortably with her inheritance and life insurance proceeds—no substitute for her late lifelong companion, but at least her few remaining years could be physically comfortable and free from the indignities of welfare and Medicare. One month of comfort was all that was granted her. As she disbelievingly surveyed the remains of her nearly emptied apartment, the radio, left behind for more valuable loot, told the story with sadistic irony in the form of the Beatles' lyrics: "He came in through the bathroom window. . . ."

In case you haven't guessed, the superintendent did it. He was an "underground locksmith" with thousands of keys and lock-picking tools. Before he installed a lock for a tenant, he made a copy of the key for future use. Some people even left keys with him in case they locked themselves out. While doing repairs he had ample opportunity to case the premises. When necessary he even made sure that a fire escape window could be easily opened from outside. He could easily observe when someone left home, and being the superintendent, his hanging around didn't arouse suspicion.

Now let us consider those windows.

3

WINDOWS

YOU SEE OUT, THEY GET IN

Although doors are the most common points of illicit entry, windows run a close second. Doors have the advantage of greater resistance to force, and windows the advantages of high visibility and difficult access. Modern residential design though has generally neglected security, and as a result almost all houses and apartments have one or more windows that can be easily—very easily—entered from the ground, a fire escape, ledge, or patio. To the criminal's delight these are often concealed from outside view, and quick entry poses little risk. Some people make an effort to secure all windows they consider accessible, then leave a ladder in the yard which can be used to reach an unsecured window. Sometimes the burglar brings his own ladder. Thieves can get through windows as easily as sunlight unless you take effective preventive action for every window you have.

THE BURGLAR'S PANE

There is greater variety in the construction of windows than of doors, but to the burglar there are only two types of windows—accessible and reachable. Some windows are less likely to be selected because of their location, but if there is only one window not secured because it is considered inaccessible, you can be sure that is the one he will try. Ladders, scaffolds, ropes, ledges, bridges made of planks—where there is a will there is a way, and the way will be found. This cannot be emphasized enough.

The window frame is the stationary part anchored to the structure, the sash is the movable part or parts, and the panes, usually glass, are mounted in the sash. The resistance to physical force of all three is often inadequate, but the sashes and panes are especially vulnerable. Frames and sashes should be made of steel, but wood, aluminum, and even vinyl plastic are the materials most often used.

Buildings are built for profit, not for the well-being of future tenants. If you are having your own home built you can control the quality of windows installed and should get the most secure, but most often you must work with what you have and must pay a high price if you desire full window replacement. Repairs, protective devices, locks, and alarms are less costly. Even the best windows should have these, especially good alarm system sensors.

The purpose of a window is to allow light and fresh air in, to give a view of the outside environment, and to permit escape in an emergency such as a fire. If a window is not usable or regularly used for these purposes, it is best to eliminate this unnecessary security risk. Boarding it up is not always as secure a measure as one might think, and is unsightly. It is advisable to remove the entire frame and replace it with a solid wall of glass blocks, bricks, or even reinforced concrete.

Translucent glass blocks permit the entry of light, and when properly installed are practically as impregnable as a brick wall. They also provide better thermal insulation and keep out noise better than ordinary glass. They are an excellent replacement for highly vulnerable basement windows and glass panes alongside or above a door frame. Glass block is not cheap and requires careful installation by a skilled craftsman, but once installed, it is trouble free, attractive, and very secure. Because it is translucent but not transparent like window glass, light can enter but your privacy cannot be violated.

As with a good door in a weak frame or a good lock on a weak door, don't worry about the fragility of window glass until the strength of frame and sash are assured. Wood deteriorates with age, especially when it is exposed to the elements, and repair or reinforcement may be necessary. If a window is frozen shut by warping or from paint, you may be better off leaving it that way if it is not necessary for ventilation or fire escape. As long as the frame is well anchored in the structure, you have a free lock that is more secure than many of the cheap, easily defeated gizmos called locks that come with most windows or sell for a dollar or two.

When an air conditioner is installed in a window it must be securely bolted to the frame or sash on the inside, preferably with angle iron and heavy roundheaded bolts or large one-way screws placed three or four inches apart. The sash must then be bolted so that even the great force of a crowbar cannot open it. This should be done even for the most inaccessible windows, before some unhired "workmen" remove the air conditioner and enter with ease.

When securing an air conditioner in a duct through the wall, make sure that it cannot be removed either by pushing in or pulling out. Do not depend on the flimsy duct grilles for protection, as they can be removed or cut easily. alarm sensors on air conditioners are advisable.

If you purchase replacement windows for thermal insulation, you should consider strength and security as important as saving energy. Are the sashes steel, heavy wood, thin aluminum, or vinyl? Are they double or triple glazed? Can they be removed from the frames from outside? Can you install good locking devices and alarm sensors in place of or in addition to the often inadequate locks they come with? Are they highly rated for security by an independent testing agency? Take your time and buy wisely; too many insulated windows can keep the cold air out but let the burglar in. More information on how to shop for these is given in a later chapter.

Glass can be a real pane. It can be skillfully cut, crudely broken, or even removed in one piece from the sash. A small opening is all that is required to open an inferior locking device or defeat an improperly installed alarm sensor. Metal foil alarm tape provides the best protection, but having an alarm does not mean that other precautions are not necessary. Aim for physical security first, then back it up with electronics. Excellent substitutes for glass are unbreakable clear polycarbonate plastic, heavy plate glass, or laminated glass with a shatter-resistant coating. The unbreakable glass is better than polycarbonate but costs more; any of them may be used in place of ordinary window glass wherever such glass is found.

A sheet of unbreakable transparent material may be used to cover the entire inside of the sash as a backup to the glass panes wherever doing so does not interfere with the functioning of the window. If the sheet is mounted directly to the sash, it should cover all the panes in it, extending at least 1-1/2 inches beyond them at sides, top, and bottom where it is fastened with many large screws or bolts placed around its perimeter. An "invisible shield" of such material is an alternative to bars or gates in securing windows that are not necessary for ventilation or fire escape. If the frames are set back several inches from the surrounding walls, a plastic sheet of the correct size can be fastened to the walls surrounding the frame. This also provides added thermal protection in cold weather, but of course the shield would have to be removable from the inside for cleaning and repair of the windows. This is no visible deterrent though, as the burglar is not aware of the barrier until he has penetrated the window; alarm tape applied to the barrier makes an excellent deterrent, as now the burglar sees that a very unusual obstacle awaits him beyond the window itself.

Installing plastic shields is a unique challenge to the home craftsman, as there are many variations of window styles, frame structures, and placement. These barriers are not limited to backing up windows alone, and the custom-made nature of such work makes it all the more effective for security.

Window glass with an embedded mesh of thin wires is superior to ordinary glass, but is not used for home windows as it is not fully transparent. It may be used for windows in a garage, the public areas of apartment buildings, basement windows, or skylights. It is inferior to unbreakable plastic or laminated glass, but is more difficult to penetrate than regular glass. However, it is just as vulnerable to cracking.

In time, putty, retaining moldings, and other materials used to secure panes deteriorate making it possible to remove an entire pane from outside with little effort. Glazier's points are available from hardware suppliers, are cheap, and are worth using when reglazing or replacing deteriorated materials in a wooden sash. They are put in before the final putty or molding is applied and make glass removal more difficult. If you use substitutes for ordinary glass panes they should be held by more secure means than are usually used for glass; if not, you are wasting your money. Consult the manufacturer or retailer of the glass substitute.

General glazing procedure:

1. TO AVOID INJURY, WEAR WORK GLOVES WHEN HANDLING GLASS. Carefully remove all old or broken glass from the sash, using a hammer, pliers, and screwdriver or chisel.
2. With the same tools, remove the old retaining clips, glazier's points, and putty.
3. Remove all traces of old materials from the sash. If the sash is metal, apply metal primer; if wood, use linseed oil or wood preservative. Let it dry.
4. Measure the exact dimensions of the sash at the place the pane is supported.
5. Get a replacement pane with dimensions about 1/8″ shorter than those measured.
6. When the pane is installed, it should only be in contact with the putty or glazing compound, not the wood or metal edges of the sash. Follow the directions for the material to be used; what follows here is a general outline.

7. Apply long strips of the glazing material to the sash groove where the pane is to be held. Work quickly so the material doesn't harden.
8. Even the material out with a putty knife, pressing it into the groove.
9. Press the pane gently into place. The material around it will be squeezed out, forming an airtight seal all around the pane.
10. Install retaining clips or, for wooden sashes, glazier's points, about every four inches or so around the pane. Points should be tapped in gently with a hammer and screwdriver. Be very careful not to break the pane.
11. Apply more compound or putty around the pane, smoothing it off with a putty knife. Do not skimp on materials or the pane will be easier for an intruder to remove.
12. After cleaning up, apply alarm tape if used (see Chapter 5). If you wish to paint the sash, allow one week for drying before doing so.

LOCKING DEVICES

Heavy, costly hardware that only prevents a window from being forced open is like a padlock on a lunchbag. Real security requires a physically independent rugged barrier as provided by bars or gates. Even if you do have such protection you should not neglect to make the window as impenetrable as possible, just as you would not do without locks on your door because you have an alarm system. The more barriers a burglar finds in his path, the more likely he is to give up and seek easier pickings. Anything less than total security is poor security.

Windows come supplied with a variety of locking devices that generally have one thing in common: uselessness. A rank amateur can quickly open or break most of them. This is also true of many window locks available in hardware outlets such as wedge locks, friction latches, and anything else that can be opened by cutting and reaching through the glass. The only devices worth getting are key-operated ones, solidly constructed, that can be mounted with much more secure hardware than the flimsy wood screws usually provided (see Figure 3-1). If possible, these should be mounted so that they cannot be seen from outside nor reached with a saw or other tool placed between the sashes. The key must be kept nearby for use in an emergency requiring quick exit through the window, but it should not be visible or reachable from outside. If the device permits locking the window open at the top (never the bottom) for ventilation, never leave more than a three-inch open space and make sure all sashes are as securely held as they would be if completely closed.

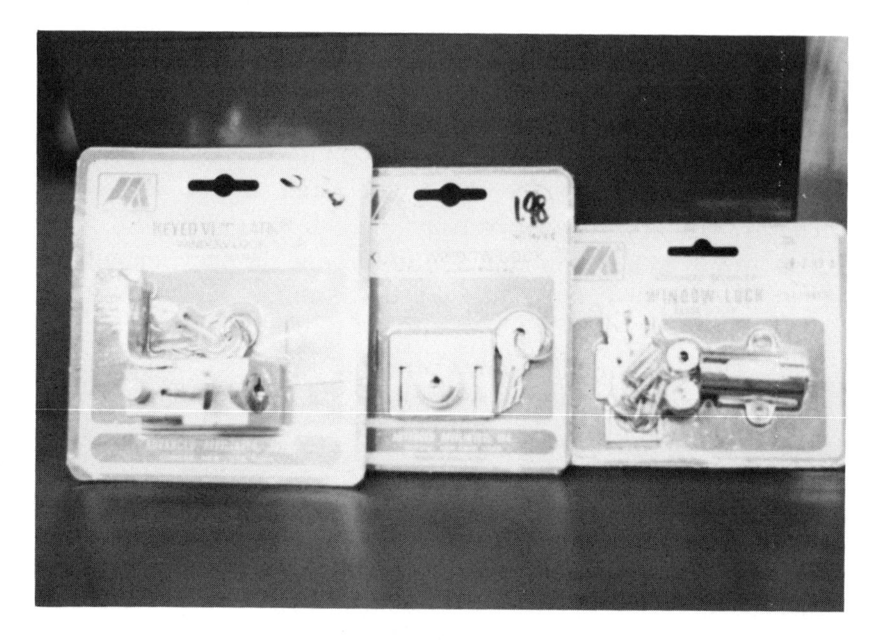

Figure 3-1. Inexpensive key-operated window locks. Usually better than non-keyed locks, but not to be relied on alone.

TYPES OF WINDOWS AND HOW TO SECURE THEM

The most common type of window is the double hung window which has two vertically sliding sashes. These are generally provided with a thumb turn lock that is very easy to break or open. Better than most commercial secondary locks available for these windows is the use of thick steel bolts that go through holes in the upper corners of the lower sash and part way into the upper sash. Use 1/4″ minimum diameter bolts; drill the holes for a tight fit completely through the lower sash and extending about two-thirds of the way through, but never completely through the upper sash. The sashes should be completely closed when this is done; if you want to be able to lock the window open an inch or two at the top, make additional holes in the upper sash that the bolts can enter when the window is open. The bolts should be long enough to extend about halfway-to-two-thirds through the upper sash when fully inserted, and the fit should be so tight that removal is difficult. Two bolts per window should be used, one on each side. If the holes are drilled for a tight fit, the security is excellent but the possibility exists that they can delay your exit in an emergency. A loose fit is more convenient to operate but more vulnerable. A variation of this technique is to drill the holes at a slightly downward angle and use thick, smooth steel pins without heads that go fully into the holes. In this way, they cannot be removed by hand through a hole in the glass. Keep a magnet out of view and reach through the window to remove the pins when you want to open the window.

Casement windows are possibly the most secure type commonly used. They should fit their frames very tightly when closed so that a prying tool cannot be forced between the frame and sash. Casement and other types of window that are operated with handles or knobs on the inside can be made more intrusion-resistant by merely removing the operating handle when not in use and placing it out of view and reach from the outside. A key-operated locking handle, if sturdy and mounted with good hardware, is even better.

Sliding windows that look like small glass patio doors may be secured in ways similar to the door. They must be protected against lifting the sash from the track as well as against opening by sliding. Several large wood or pan-head screws should be inserted in the upper track and protrude far enough to prevent removal of the sash from the frame. Placing a loose-fitting dowel or pipe in the lower track to prevent sliding the window open may not be adequate, as it often can be removed with a tool from outside. Bolts going through holes in the tracks and into the sash are good if they cannot be seen and removed from outside. Key-operated locks, properly installed, are also useful.

Louvered jalousie windows are highly vulnerable and difficult to strengthen. They can be protected to some degree with gratings or bars, but are best replaced. If you are buying a home or apartment, make sure that you do not get stuck with these; let the owner replace them or deduct the cost of your doing so from the selling price.

Picture windows or other windows with large panes are vulnerable to entry by glass cutting, breakage, or removal. Better substitutes for ordinary window glass are a must for these.

Windows that are hinged at the bottom and open outward from the top (or vice versa) can be made more secure by using stronger hinges with better fastening hardware. The best hinge to use is a piano hinge that runs the entire width of the window. Be sure to use a key-operated lock in place of the regular latch as well.

All types of basement windows are highly vulnerable because of their location. Replace them with bricks or glass blocks or secure them with thick, well-anchored steel bars.

What is true of doors is true of all windows; if they are not closed and locked, even a child can enter with ease. A burglar may be temporarily discouraged by a well-secured residence, but may keep watching for the one time a window is left unlocked. Don't give him the chance to strike.

DO BARS A PRISON MAKE?

Unfortunately, those of us living in or near high-crime areas often must have bars on our windows for lack of bars surrounding the criminals that prey upon us. Such highly visible security measures should be employed with discretion, as they can make one residence conspicuous among surrounding ones and thus tempt the skilled professional.

Just as there are ways through the best locks and doors, there are ways through the best windows. Given the choice between backing up a window with either a good alarm system or additional hardware, the alarm is the better choice. If an intruder has the skill and is willing to spend the time to get through a window in spite of a visible gate, he probably can get through the gate too. He knows what to expect and is prepared to handle it. Bypassing an alarm is a whole different ball game, and for many burglars, the sight of alarm tape on the panes is a greater deterrent than ordinary gates or bars. Chapter 5 covers alarms in depth; now we take a look at the hardware approach.

Want to play Superman? Get a long strip of thin flat aluminum that is wide enough to look sturdy. If it were steel, bending it with your bare hands would be a challenge, but with aluminum you can easily play the strong man. Aluminum is great for many purposes, but not security. It is easily bent, ripped, drilled, and sawed, and any window gates, bars or grilles made of it provide only illusory protection. If it's not good steel, it's not for real.

Iron bars and grilles that are permanently mounted outside a window may loosen in time because of corrosion or deterioration of the frame or masonry. They should be tested with a crowbar, using plenty of force; it is better to discover the weakness and repair it yourself than to let a burglar discover it. Badly rusted bars or grilles and rotted frames or masonry must be replaced, and new or slightly rusted ones should be cleaned and painted to prevent future deterioration. Bars are preferable to grilles, some of which are not as strong as one might think.

Permanently installed outside bars afford excellent protection. The best bars are made of heavy welded steel and are deeply anchored in cement or brick in many places (see Figure 3-2). Unfortunately, the necessity for a means of escape in the event of fire makes the use of these on all windows unwise, and, in many areas, illegal. Should one window be left unbarred for escape purposes, it should be the one that is the most visible from the street and nearby homes. It should be well illuminated at all times and secured with an interior gate that permits easy escape in an emergency.

Outside grilles that are secured with padlocks can be no more secure than the locks and hasps. They will protect your windows from a stray baseball, but not from a competent burglar. Decorative ironwork, whether used for protecting windows or doors, or used as a fence, may lack sufficient strength, so be sure that security is your main criterion in selecting such hardware.

Figure 3-2. Steel bars and framework here protect an otherwise highly vulnerable window and air conditioner.

Accordion (or scissors) folding gates and similar devices are for inside use only. They must be chosen with care, as some are inadequate; consult the reports of independent testing agencies such as *Consumer Reports* before buying. No matter how strong and foolproof a gate may be, it is practically useless if it is not well anchored to a solid structure on all sides. Attaching a gate to a deteriorated frame is foolish; replace or reinforce the frame first. It is best to extend the gate bolts into the masonry behind the frame. Small wood screws or nails are inadequate for this job, no matter how many are used. Accordion gates should run in steel channels on the top and bottom, and have a locking mechanism that is impossible to open by reaching through from the outside. If the window is not necessary for fire escape, the gate may be secured with good padlocks. Keys should be located in plain view near the gate, but not where they are visible or reachable from the outside. In many areas, gates with key-operated locks on fire escape windows are prohibited by law (see Figure 3-3). Ultimately, whether or not to compromise one form of safety for another is your decision.

Window guards that prevent a child from falling out of an open window do not protect against illegal entry and should never be relied on for that purpose.

Anchoring to stone, concrete, or solid brick is not difficult when you have the proper tools and materials. Use threaded steel bolts of 1/4-inch minimum diameter, and of such length that at least three inches can be embedded in the masonry. Carefully mark where they are to be placed, using the gate, bars, or whatever is to be mounted as a template. Have several nuts that fit the bolts, two per bolt, and use a masonry drill to make holes large enough in diameter to accept the bolts with nuts on them; for a 1/4-inch bolt, a 1/2-inch hole will usually be adequate. Drill the holes to a slightly greater depth than the bolts will project. Pass a bolt through each mounting hole of the gate or bars, and put two nuts on each, at positions about 1/3 and 2/3 of the length of the shaft. Use cement to secure them in the holes. Put the gate or bars in place so the cement fills the holes and completely surrounds the nuts and bolts. When the cement has completely hardened, weld the heads of the bolts to the steel frame of the gate or bars. With many bolts used, this is a permanent installation that will resist tremendous force. If you lack the proper tools or feel too inexperienced to do the job well, let a carefully chosen professional do it for you.

If you find that bars and gates are too unattractive, consider using solid wooden shutters with secure hinges and locking devices on the inside. However, these may not be as resistant to force, and an alarm system with window sensors is advisable. Roll-up steel shutters are unattractive, but are comparable to bars if not better in the protection they provide. They should be used on garages, worksheds, and high-risk windows, such as those facing into an alleyway. Shutters must be well-anchored and should lock from the inside, as outside locks are more vulnerable. Check your Yellow Pages under "Gates," "Shutters," "Iron Work," and "Guards—Door & Window" for local retailers.

It is most practical to hire a contractor for major projects, such as putting in more secure windows or installing gates, shutters, or bars.

Figure 3-3. A fire department-approved window gate. The small box on the right contains a lever with which the gate can be quickly opened in case of fire.

As in other areas of home repair, there are some that are less than reputable. Here are some shopping rules:

1. Deal only with an established, reputable company, *never* a door-to-door salesperson who offers low prices. Be suspicious of high-pressure advertisers seen in the media; they have to either charge you more or give you less in order to pay for all their expensive hype.

2. Visit several stores/showrooms and ask questions before making any purchases. Examine the available products and get information on their merits and shortcomings regarding intrusion resistance, thermal insulation, and so forth. In evaluating windows, consider how easily and effectively you can install better locks and alarm sensors. Be sure to visit many retailers; you must compare the biased opinions of salespeople and decide what is best for you. It pays to take your time when making a major purchase that you will use for many years to come.

3. When making your purchase, get a detailed, written contract. As some windows, shutters, and gates have no brand name, the contract should specify the frame materials, type of glazing, weather-stripping, and other details. The contract must also give the total price, including installation, and the terms of the warranty on materials and labor.

4. When your purchase arrives, carefully check it *before installation begins* to make sure you are getting what you want. If the wrong materials are sent, do not allow work to begin until the right materials arrive.

5. Check all work carefully before the installers leave.

In addition to having well-secured windows, you should minimize their accessibility wherever possible. "Cat burglars" are not pet thieves, but strong, agile, usually young males who climb like monkeys to reach points of entrance that most people consider inaccessible. This is not the modus operandi of most burglars, but of a large enough number to make preventive measures worthwhile. Any window is accessible to a determined burglar, so make his job as difficult as possible. Burglars, particularly amateurs looking for a fast buck, take the path of least resistance. Resist! Cut off any tree branches that can be climbed to reach a window ledge, patio, or roof. Use iron grillework or bars to block travel on and access to ledges. Wherever possible, remove outer windowsills and other structures upon which someone can stand or climb. Fire escape ladders must be suspended as high above the ground as possible. To prevent access to fire escapes, the roof doors of apartment buildings should be kept locked from inside with alarm locks: these may be opened in an emergency by pressing a bar that then sounds an alarm. They are not foolproof and the battery must be replaced routinely, but they make access to the roof more difficult. For legitimate use of the roof, the lock may be opened with a key without triggering the alarm. An alarm lock also may be used in place of a hand-operated sliding deadbolt to secure fire exits in apartment and public buildings (see Figure 3-4).

Given a choice between two similar windows, the burglar will try the one where he is least likely to be observed. If he is certain he is well concealed, he will spend a good deal of time and effort to break in. For this reason, windows that are concealed by shrubbery, enclosed porches, surrounding structure such as an alley, or hidden by darkness are likely targets. The answer is to eliminate these obstacles to visibility as much as possible and to provide good illumination during all hours of darkness with lights mounted where they cannot be tampered with. Such lights can be controlled automatically by a photocell relay device; at least two lights should be used in case one burns out. Thieves may be good actors in court, but they hate to work in the spotlight.

Figure 3-4. An alarm lock. Not high security, but better than a hand-operated sliding bolt for roof and fire exit doors.

4

OTHER WAYS

"Twenty-three hundred dollars! Twenty-three hundred dollars—and look! My God, I'll kill that bastard who sold us this junk, right after I kill the rat that did this to us." Archie was angry, a condition that did not bring out his better nature. "I toldja we shoulda moved" wife Tiffy added in a whining, sarcastic tone. "The Milktoasts moved out a year ago. Said the neighborhood waz bad." All Archie could do was stare at the mess before him. $2300 for locks, window gates and an alarm system, and yet the house was burglarized during the Gull family's first vacation in three years. "Maybe it ain't what it looks like, Archie, maybe Jejune came home fer a day or two an wid some friends she messed things up." "Like Hell—her wimp friends couldn't mess up a dollhouse. Call the cops! Call the insurance company! Lucky they didn't get my 38, cause if some creep shows his face near here I'm blowin' him away! Well get on the phone, woman!"

Five hours and two calls later, Inspector Onus Probandi and Officer Intrepid of the central precinct arrived. "Are you the Gulls?" Intrepid spoke with reassuring authority. "We sure are," Archie replied with a very uncharacteristic lack of respect, "we been robbed, and the insurance company got here faster than you!" "Sorry sir, but we had several emergency calls." As the men stepped up the porch, shrieks were heard from the kitchen. "Archie, the plates is gone! They took the plates! Hardly anything else's gone but they got the plates! Oh God!" "Jesus! Those plates was collectors items—we paid a fortune for 'em, almost as much as for the damned security system that didn't do crap for us." Archie's anger was now reduced to a pathetic gloom.

After reassuring the Gulls that they would do all in their power to apprehend the thief and recover their property, Intrepid and Probandi were introduced to Mr. Feasel, claims adjuster for the firm of Welsh & Co. Feasel had been fairly quiet up to this point, but now spoke sharply. "I'm glad to see you gentlemen, but I'm afraid this may be a waste of your time. As far as I can determine there has been no larceny here; no evidence of forced entry, and a perfectly good alarm system which these people have stated was in use when the alleged crime occurred." After stepping between Archie and Feasel to prevent another crime, the officers inspected the premises and listed the things allegedly stolen. Many large items such as the color TV, air conditioner, and stereo had been left on the living room floor but not taken. "Couldn't get 'em out without opening a door and setting off the alarm?" Probandi inquired of Intrepid. "Then how did they get in?" interrupted Feasel. "I'm afraid I cannot stay any longer, gentlemen. My report will show that no clear evidence of breaking and entering has been found. If you find any evidence to the contrary, either you or the Gulls should call me. Meanwhile, I will contact the Huckstar Security Company, who I'm sure will confirm the integrity of their alarm system and the unlikelihood of this incident being a burglary as is alleged. Good day." At his departure Archie uttered several obscenities rarely heard even by the experienced ears of the police officers; Intrepid in fact was unsure of the meaning of one but declined to inquire. Tiffy timidly said "oh my" and scurried off to the kitchen with her hand beside her face to hide her embarrassment. Archie shouted at the cops, "Well, ain't you guys gonna look for prints or something?" "I'm sorry, sir, but standard procedure only permits such investigation in clearcut cases of grand larceny with a loss in excess of five thousand dollars. We will provide you with a report number for your insurance company and for claiming a theft loss as an income tax deduction." "Ha," Archie shot back, "the insurance bastards won't pay a cent and the IRS is doing away with that deduction. Nobody wins but the crooks!"

In the week that followed, Archie called the precinct almost daily with no satisfaction. Ten days after the discovery of the crime he finally went to see the Inspector in person. "Probandi! You've been in this thing for over a week and nothing has come of it yet! The insurance company won't pay if I can't prove it was a real burglary." With patience and self control the Inspector answered, "Mr. Gull, I can understand your being upset and I would help if I could, but I typically put in a 45 to 50 hour week working exclusively on cases of homicide, kidnapping, and other felonies. Have you done any detective work on your own? Have you asked your neighbors if they saw anything suspicious, or if they have been burglarized recently?

This is not that big a city; the culprit may very well live right down the block from you. If you can find us a lead, we will follow up on it, I promise."

Upon arriving home Archie told Tiffy to speak to their neighbor Frank, who being a notorious blabbermouth would be sure to tell if he had seen any prowlers in the vicinity. "But Archie, you said you didn't want anybody to know. He'll tell everyone an' we'll look like fools. The biggest gossip in town has got to be Frank Lipschitz." "Don't argue, woman, do as you're told. If he seen anything we gotta know about it."

Tiffy returned in ten minutes with Frank. "Hey, Archie, you should've said something sooner. I saw them two teenage girls that hang out by Ryan's foolin' around by your house one night, but when you got home and didn't say nothin' I thought everything was OK. It looked like they were pushing on the wire grating over your air conditioner duct under the side window." "You jerk! You shoulda called the cops right then! I'm calling that Inspector and you're telling him all you know. Jesus, girls! Who would believe it!"

Probandi made good on his promise, and the girls were busted. The valuables were not recovered, but with the confessions of Ms. Ann Thrope and her accomplice Rapine the insurance company was forced to pay for their replacement. Together the girls had removed the grating and pushed the air conditioner into the room, then entered through the duct it had been placed into without secure fastenings. The alarm was not connected to it, so did not sound. They left by the same route, replacing the grating before fleeing.

Being first offenders (translation: first time caught) Ann and Rapine were released on their own recognizance. Being well able to afford a good lawyer, the girls got off with probation.

Moral: Watch your air conditioners!

LIGHTS, FENCES, AND WARNINGS

A criminal's fear of being seen and caught in the act is a powerful deterrent. For this reason, good illumination at night and the elimination of structures or landscape features that provide concealment are necessary. These points have been discussed in regard to windows, but apply to all possible entry points and nearby areas such as front and rear yards, porches, patios, driveways, garages, fire escapes, fire exits, alleyways, and recreation areas. Lights should not be aimed at an area or house from the ground because these are vulnerable to tampering and the glare may interfere with your view from the windows. It is best to mount lights under the eaves or other elevated, hard-to-reach places, with the wiring run so as to be inaccessible from outside.

Lighting should be uniform and positioned to eliminate shadows rather than to cause them. As well as being out of reach to prevent tampering, fixtures should be aimed so that the direct glare of the light source is not blinding to anyone viewing the illuminated area from outside. Fixtures should be easily adjustable for aim, weather-proof, and reflectorized for efficiency. When replacing a bulb, be sure to clean the reflector and glass covering so that efficiency is maintained.

Fluorescent lights are preferable to incandescent lights as they are more efficient and cost less to operate for the same amount of illumination. They also need to be replaced less often. When purchasing lights, be sure to specify fixtures made for outdoor use, even if they will be shielded from the elements. Mercury vapor lamps are also efficient sources of intense illumination. Sodium vapor lamps are efficient and very intense, but their yellow-orange glare is somewhat unattractive. Whatever form of lighting you use, it is best to have a photoelectric control that automatically turns the lights on at dusk and off at dawn. This will always operate the same way whether you are at home or away, and assures adequate illumination at all times.

How much light do you need? Again, the principle of a low profile applies. A bank of 300-watt bulbs under the eaves will illuminate your yard like a night game at a baseball stadium, but aside from the high cost of power consumed, they will also tell a potential burglar that you are obsessed with security, therefore you have property worth taking. If the burglar is highly skilled, the lights will not stop him. A uniform level of illumination that is slightly greater than that existing in all surrounding streets and properties is sufficient. You may wish to spotlight possible entry points such as driveways, side or rear doors, and windows, but do so with discretion. However, vulnerable points in high-crime areas, such as apartment building fire escapes, require maximum illumination.

The installation and maintenance of street lights is the responsibility of local government, and you should insist on quick repair when a light is out. Likewise, if you are a tenant you should insist on immediate repair of any defective interior or exterior building lights. If any lights have been vandalized, be on guard as some criminals do this in preparation for their work.

An interesting gadget on the market is a lighting device that mounts on the exterior of one's home, well above the ground, and automatically illuminates the area at night when it senses someone or something moving within its range. This uses a passive infrared sensing system, and the light automatically goes off several minutes after the disturbance ceases. It should not be used in place of good nighttime illumination, but as an added deterrent. Prowlers quickly get the message that you are security conscious and that perhaps they should go elsewhere.

While it is desirable that your home and surrounding land be highly visible to nearby neighbors, it is not a good idea to make a show of affluence to the public. A luxurious home that is visible from a nearby highway sits like a plump hen beneath the flight path of chicken hawks. An ambitious predator can descend quickly, seize its prey, and be off and lost in the crowd within minutes. Landscape architects should consider security as well as aesthetics, applying the principle of a low profile. A row of trees may be used to conceal a home, garage, or swimming pool from view from a hill or a roadway. What others can't see can't tempt them.

Fences, like window bars, are ugly and isolating, yet they have a role in protecting us from the evil in our midst. The average fence poses no barrier to someone truly determined to get over, under, or through it, but it can discourage some trespassers and is another element in your arsenal of deterrents to criminals. If you get a fence, it should be one that does not block the view of your property and thereby provide concealment for any intruder who gets past it. A four or five-foot high chain link fence with a gate, secured with a good lock, is adequate. Such fences are available with smooth tops that are easy to climb over, or tops with protruding sharp wire ends that help to deter climbing. There may be a local ordinance that limits the size and type of fence that may be used around your property. Such laws should be abolished, but be aware of the possibility of trouble should you violate one.

Iron bar fences can be sturdier and more ornate than chain link, but vary in their effectiveness because of design. A fence that is made of vertical posts and smooth, regularly spaced crossbars is practically useless no matter how high it may be as it can be climbed like a stepladder. The best test of any particular fence design is to have an agile, athletic person try to climb it. If he finds it very difficult, and the fence is solidly built and welded, it is worth having.

Even high barbed-wire or razor-edged-coil-topped fences can be bypassed. They can be cut through, and in time the sharp points rust and become ineffective. They are unwise for the same reason that

"Beware of Dog" and other common conspicuous signs are unwise—they ask for trouble and advertise that you have property worth stealing. Signs that you should have, however, are large, clear, well-illuminated ones with your address. One (on which your house or building number is sufficient) goes on the front, the other, with your full address, goes on the back. Anyone who sees a prowler around your home can now tell the police the exact address to go to, and when they arrive, they will see your home immediately.

Warning signs and stickers are of limited value; they deter some amateurs but will not fool the smart crook. Visible evidence of genuine security precautions such as alarm tape on all windows is a more effective deterrent to the skilled, experienced burglar, as well as the amateur. Common, mass-produced warning signs have lost their psychological impact and are generally considered to be no more than a bluff. If you want to use signs, they should be custom made—easy to do yourself with a typewriter, felt-tipped pen, or paint. They should not be too large or conspicuous, but be placed where an intruder will be sure to see them before making his attempt. The wording should be concise, credible, and calculated to discourage.

COVERING ALL POSSIBILITIES

We take for granted that our walls, floors, roofs, ceilings, and so forth are impervious, but this is not so. The ordinary burglar will not take the trouble and risk, but a professional with his eye on a bonanza will smash through concrete to get to it. Skillful pros have spent entire weekends tunneling into safe deposit vaults in banks and have made off with millions of dollars without triggering an alarm. This only demonstrates the lack of imagination of the designers and installers of the alarm or an unwillingness on the part of the bank management to spend enough on security to be really thorough. In the next chapter, alarms are discussed in depth, including how to protect walls and other surfaces.

Often, very small openings such as milk or coal delivery chutes, "doggie doors," and chimneys are considered impossible entryways, but there are burglars of small size or with children as accomplices who can practically slip through a soda straw. Some have even been known to conceal themselves in a large carton along with merchandise to be delivered to a home or business, to emerge when people have left or gone to sleep and proceed with a burglary. Beware of "Santa Claus in reverse"; he may show up anytime. Skylights, elevator shaft doors, dumbwaiters, chutes, chimneys, air vents, fans, air conditioning ducts, attics—all of these must be both well-secured physically and wired into your alarm system. Make no exceptions.

Great care must be exercised in the selection of building contractors, babysitters, or others who work for you occasionally. As good as the company's or agency's reputation may be it is possible that, unknown to them, one or more of their employees is dishonest. They may "case" your home for a future "visit," sabotage your alarm or other security equipment, or grab whatever they can while your back is turned. Workers should be watched closely, and if you cannot do this personally with at least one other person helping you, it may pay to hire reputable security guards for the period the work is in progress. Well before the workers arrive make sure that all valuables and keys are removed from sight or from the premises entirely, and when the work is completed, be sure to take an inventory of these items. Also, check your alarm system and locks thoroughly. If you are having wiring or new construction done, see to it that incoming power and telephone lines are run underground or otherwise made inaccessible from the outside.

When moving, be sure to provide good security in your new home before leaving any of your possessions unattended there. Your most valuable and delicate belongings are best transported yourself; the rest must be well packed in boxes identified only by a code on the outside, such as "L2" for living room, box 2. Keep a written inventory list in your possession, and before the movers leave, check that all boxes are present and have not been tampered with. If anything is not as it should be, call the police.

Should some disaster such as a fire or storm ever render your home or apartment uninhabitable, *get all the property you can salvage out as soon as possible*. The police may claim that the premises will be watched, but get it out anyway; it is your property to protect as you see fit. Looters are as uncaring as the vultures that hover over the dying man in the desert, and now and then one even wears a police or fire uniform.

Baby sitters should be relatives or good, trustworthy friends. Keep in mind though that good people may have bad "friends" who, unknown to them, take advantage of the relationship for illegitimate purposes. Teenage girls often invite boyfriends over for awhile when baby sitting, and even conservative middle-aged women may be or know people with sticky fingers. Before hiring, thoroughly check anyone you do not know well. Baby sitters should be told not to open the door to strangers or have any guests without your consent, to call the police in the event of trouble, and to call you (leave the number where you can be reached) if a nonemergency problem arises. The phone should be off the hook unless you want to be able to call to check on things; in that case, caution the baby sitter never to admit to a caller that you are not home, but to take the number so you can call back. The sitter can then call you and relay the message. The sitter should be told, if necessary, of any hazards in the house and important security precautions, but should never be given information you would not want known to strangers. If you find any of your instructions have been disregarded, never employ the person again. Make sure when you leave that all doors and windows are locked and that outside lights are on, and be sure to call and notify the sitter if you will be home later than expected.

Although most people take great care in finding a reliable baby sitter, few exercise the same care in protecting their property. A good private security guard service is hard to find. According to a New York State Investigation Commission report based on 1980 figures, about half of the private guards in the state have criminal records, and many private security companies are merely fronts to obtain gun licenses for unqualified persons. The report states that some firms engage "in the very criminal acts they were hired to prevent," such as burglary and vandalism. New York City police investigations into forty such firms have shown that 65 percent are "delinquent in one form or another," and many went out of business as soon as investigation began. If you need a guard service for a short time, you may be better off hiring some good friends with legal weapons or an off-duty cop or two. If you want a reputable security firm for long-term employment, check them out very carefully with the police, Better Business Bureau, and consumer protection agencies. A good firm should have several years of experience under the same company name and be able to provide many references. Check them all.

COVERING IMPOSSIBILITIES

Many sources of information on burglary prevention take the defeatist attitude that you will become a victim in spite of precautions. They give almost as much information on what to do when the crime is discovered as they do on prevention. Negative thinking brings negative results; that will not be done here. In brief, never enter a residence when there is a possibility the criminal is still inside. *Immediately* call the police from the nearest phone and tell them there is a "burglary in progress" at (full address). If the burglar is gone, do not disturb anything, but call the police, and then your insurance company if you are insured. Be aware that unless there is evidence of forced entry you probably will be denied payment under your policy; see Chapter 10 for important information on this.

If you are awakened by the sounds of an intruder, first lock your bedroom door, and then hit the panic button if you have an alarm, although a good alarm properly used would have already sounded. Quietly call the police, give them the full address, and tell them to hurry as you may be attacked at any minute. What to do next depends upon how you have prepared for such a situation and how much risk you are willing to take. If your alarm is sounding, chances are the burglar is running like crazy and you are safe. Three things you should have within reach of your bed are the panic button, a phone, and, if you own one, your gun.

It is *very dangerous* to leave your bedroom if an intruder may already be inside your home. Even if you are armed, you are at a strategic disadvantage. You are alone, drowsy, and are an exposed target; he or they are alert and may be in ambush positions, waiting for you. Once you have called the police, it is *their* fight, and even if there is no burglar, you could get shot by a policeman who mistakes you for one. Take cover, and if a criminal breaks into your room, act accordingly, but DO NOT go out looking for a battle. You will probably lose.

An unarmed woman alone, hearing an intruder outside the bedroom, can employ psychological trickery for defense. Loudly say something like "Jim, grab the gun, someone's in the house!" The intruder will probably flee, or proceed very cautiously, during which time you can call the police.

Should you unexpectedly confront an intruder face to face, remain calm. He may be armed, and although you may want to wring the creep's neck, it would be dangerous to attempt it now. Do all you can to prevent violence; do not reach for a weapon, make any sudden moves, or say anything threatening or provocative. Unless it seems to disturb him, maintain eye contact and calm, relevant conversation. Make a mental note of his physical description. Speak deferentially; tell him you will not resist and that he may take anything he wants. Chances are he is as scared as you are and will leave quickly, in which case let him. It is safest not to fight unless you are attacked, but if you must, disable or kill as quickly as possible.

The situation of an attempted breakin while you are at home and awake was discussed in Chapter 1. Criminals who would brazenly enter in your presence, as by smashing in through a door or window, must be assumed to be armed and dangerous and must be handled accordingly.

COPS AND ROBBERS AGAIN

This game, as described at the end of Chapter 1, is worth playing again from time to time, especially if you have moved or made other changes in your living space. New ideas will always come up, and as each is acted upon your home and family will become more safe and secure.

FOREWARNED IS FOREARMED

One of the most emotionally debated crime-related issues is that of the possession of a firearm for purposes of self-defense. It would be a safer world if guns and other weapons did not exist, but they do, and we must face that reality and deal with it intelligently.

Roughly one-half of all American homes contain one or more guns, and approximately 150 million firearms of all types exist in the United States. A 1980 study by the Federal Bureau of Investigation found that of 21,860 murders in one year nationwide, 62 percent were committed with guns. Among these the handgun was by far the most often used, with only 12 of the 62 percent being committed with a long gun (rifle or shotgun). Knives were used in 19 percent of all murders, rifles in 5 percent, and shotguns in 7 percent. Of all gun owners, about 70 percent have rifles, 66 percent have shotguns, and 49 percent own handguns. According to the Consumer Product Safety Commission, there are about 26,000 gun-related accidental injuries annually, one third of which require some hospitalization. The National Safety Council reports approximately 1,900 annual firearms accident fatalities.

While many states and localities have enacted more stringent gun control laws, and these vary widely, the 1968 Federal Gun Control Act regulates nationally the manufacture, import, and sale of all firearms. The local laws are directed mainly against the handgun, the weapon most often used for criminal purposes. Under Federal Gun Control Act regulations, no person under 21 years of age may purchase a handgun, and the following individuals are prohibited from owning firearms:

1. Any person under indictment for or convicted of a crime punishable by over one year in prison.
2. Any fugitive from justice.
3. Any person "addicted to marijuana or narcotics."

4. Any person who has been judged mentally defective or has been confined in a mental institution.
5. Any person discharged dishonorably from the military.
6. Any person who has renounced his or her citizenship or is an illegal alien.

The interpretation of the Second Amendment of the United States Constitution is a point of dispute between those who oppose and those who favor gun control. Opponents contend that private ownership of weapons is their constitutional right, control advocates say that the amendment refers to weapons in the hands of the militia only. Supreme Court decisions have tended to favor the latter interpretation.

Advocates of stricter gun controls wisely aim their efforts primarily at the handgun, but all do not agree on what specific measures should be taken. They cite crime statistics as evidence that the 1968 Federal Gun Control Act is insufficient and that state and local laws are too varied and inadequately enforced. They point to statistics that indicate that a gun kept for protection is more likely to kill someone you know than an attacking stranger. The statistics claim that 90 percent of all burglaries take place when the residents are out, with the gun usually being stolen for future criminal use. Murder rates are lower in countries with strict uniform gun laws, they point out, and the majority of handgun murders are not planned felonies but crimes of passion. Street criminals rely on speed and surprise; unless intended victims are very alert, they may not get a chance to use a gun for defense and may wind up having it used against them. The American Civil Liberties Union states that the dangers of guns justify their regulation. The ACLU supports this interpretation of the Second Amendment, but concedes that certain aspects of regulatory laws and their enforcement may violate individual rights of privacy and civil liberties. It is undeniable that a gun accident cannot occur without a gun present, nor can a nonexistent gun be seized by an intoxicated, crazed, or criminal individual for illegal use.

Gun freedom advocates, on the other hand, point to lower crime rates where citizens possess guns and are trained in their use. They say that statistics show that gun control does nothing to reduce crime and violence. New York City, with very strict gun control laws, has a high incidence of crime and firearms violence; in Orlando, Florida, the crime rate fell by over 50 percent after a police-administered public course in self-defense with guns. Opponents of gun control seek stronger measures against criminals who use guns, not against the guns themselves. They counter the foreign crime rate argument with the fact that the incidence of all crime there is lower, not just crimes involving firearms.

The vast majority of gun owners are sensible and law abiding. A very small core of criminal repeat offenders, possessing illegal guns, commit most of the crimes that are attributed by control advocates to the availability of legal weapons to the general public. Opponents of further gun restrictions feel that public ignorance, not the inherent nature of guns, is the chief cause of accidents. Education is required in the safe use and storage of weapons, not bans by government fiat that in effect treat the entire public like untrustworthy children. Privately owned weapons not registered with the government are considered by some to be an excellent guarantee against possible future dictatorial suppression of the people.

Although perhaps only 10 percent of all burglaries occur when someone is home, it stands to reason that these, along with brazen intrusions by armed robbers, would decrease with the increased probability of meeting armed resistance. National Crime Survey data indicate that the use of a weapon for self-defense decreases the chances of the criminal's success and also decreases the likelihood of injury to the victim. In robbery cases, for example, the victim was injured in about 14 percent of the incidents where a gun or knife was employed for defense, as compared to roughly 33 percent when no means of defense or resistance was used. As to whether or not the benefits outweigh the dangers of gun ownership, the 1978 Caddell survey (Cambridge Reports, Inc., commissioned by a pro-gun control group) found that 3 percent of the American adult population had used a handgun for protection from a criminal, as compared to 2 percent who had been involved in a handgun-related accidental injury. In a summary of findings of two national surveys in 1978 by Decision/Making/Information®, commissioned by the National Rifle Association, it was found that self-defense is the strongest motive for gun ownership. In addition, 7 percent of the respondents (or 6.6 million Americans) said that they or a family member had used a gun for self- or property-protection against another person. Nine percent said they had been in a situation where a gun was needed for self-defense but none was available. The report states that as they are based on the memory of those surveyed, these percentages may be an understatement. Of course, those that did not survive their encounter with a criminal could not respond.

Statistics can be expressed in such ways as to appear to support both positions, as both sides of the gun control issue present impressive figures in their favor. Ultimately, people accept the statistics that support their own viewpoints. This is an issue calling for rational judgment and a careful weighing of the facts.

Although my views are liberal on many issues, they do not conform to the liberal positions on gun control and certain aspects of the handling of the crime problem. The savage assassinations of Martin Luther King Jr. and Robert Kennedy grieved me; I was even more deeply moved by the senseless murder of John Lennon at the hands of a gun wielding maniac. With these courageous people died some of our hopes for a saner, more peaceful world. They were not killed by guns however, but by malicious people using them. Had these victims been armed, or at least had they been protected by alert armed bodyguards, they might still be among us. The world is worse for their loss, and may have been better had a gun been in the right hands at the right moment.

A recent experience of mine illustrates how the presence (or rather the imagined presence) of a gun can actually prevent crime and violence. As I approached the ramp of a footbridge over a highway, a suspicious looking youth passed me slowly on a bicycle, mumbling a snide remark. Reeking of "bad vibes," he put me on immediate alert. Another bicyclist passed slowly, saying nothing, as I was halfway up the long ramp that had fences on both sides. The two teen-agers stopped in the middle of the bridge, pretending to enjoy the view. My suspicion of danger was confirmed when, looking back down the ramp, I saw that a third savage—and I make no apologies for the word—was blocking my retreat. Confidently, I put my right hand in my pocket, feigning the presence of a gun, and walked back down the ramp to confront the single would-be attacker blocking my path. He stood aside. Looking back, I saw all three fleeing over the bridge. In later conversation with a police officer, I learned that there had been recent muggings done that way at the bridge.

I believe it is totally naive to expect laws to keep guns out of the hands of the lawless. The tighter the controls, the larger the black market. The problem is not so much guns as it is ignorance, criminality, and insanity, and these are what must be remedied. The criminal's tool should not be made the scapegoat for our frustration in failing to achieve the rehabilitation of the criminal's mind. That same tool in the proper hands can save lives, as can the knife in the hands of the surgeon. Where strict gun control (translation: gun banning) prevails, guns are only carried by criminals, an inadequate number of police, and some otherwise law-abiding citizens who value their survival and rights to self-defense enough to risk possessing an illegal weapon. Guns, like autos, should not be in the hands of the unfit or untrained, but if all honest citizens are disarmed and police protection is inadequate, what choice have we but to submit to the mugger or rapist and helplessly beg that our lives not be taken? The criminal is safely brazen when sure that his intended victim and all bystanders are unarmed. In a sane and just society should it be the decent individual or the criminal that must cower? Should we or they be looking up the barrel? As American citizens with the Constitutional guarantee of domestic tranquility we have the RIGHT to walk the streets and use public transportation without fear; the RIGHT to the safety and security of our persons and possessions. We have been negligent in surrendering these rights; we must take them back, as non-violently as possible. Personally, I would prefer to face charges of weapons possession than lie six feet beneath an epitaph which reads "LAW ABIDING TO THE END." Just as there is danger in our means of national defense against foreign attack, so there is danger in civil self-defense. I would prefer living with that danger to surrendering my freedom to the uncivilized among us. Survival is the first law of nature; let the laws of fallible men not subvert it!

A lethal weapon in the hand or home of an unqualified individual is highly dangerous. You should consider gun ownership only if you are certain that you can make rational decisions while under stress, are willing and able to keep the gun out of the wrong hands, and learn its proper care and use. You as owner will be fully responsible for the weapon.

Rifles are not ideal for home defense. They require careful aim; if you miss on the first shot, you may not get a second chance. The bullets penetrate walls easily and may injure someone at a considerable distance. The long barrel makes a rifle impossible to conceal and a bit clumsy to handle; an intruder can disarm you by unexpectedly grabbing the barrel as you move around a corner with it. Save the rifle for sporting use.

Handguns, although statistically the most dangerous of firearms, are usable for self-defense and home protection, particularly when the owner is well practiced in the use of his or her particular weapon. The handgun is the easiest of guns to handle, it is concealable, compact, easily stored, and ready for quick use. Some experts consider a good handgun with the proper ammunition to be the best of all home defense weapons, although it is the easiest weapon to lose to theft. In most localities, however, restrictions on handgun ownership are more stringent than on any other weapon.

While small-caliber guns such as the .22 are fine for learning and target shooting, they are inadequate for self-defense. Even larger bore guns with ordinary ammunition, such as roundnosed slugs, can lack sufficient stopping power. A viciously determined, crazed, or drugged assailant can take several hits with these and still keep attacking. For defense you want as effective a firearm as you can learn to handle. Recommended are double-action revolvers (single-action ones are too slow-firing) and single-action automatic pistols of .357, .38, and 9mm. caliber. Larger guns, such as the .45, pack a wallop (recoil) that some find hard to control. For ammunition, use only hollowpoints or Hydra-Shoks. The .38 Special +P and 9mm. rounds (bullets) are good for defense.

Shotguns are good home defense weapons, although they have some of the same drawbacks as rifles, primarily their size. Their main advantage is the spreading out of the shot pattern upon leaving the barrel, making precise aim less critical, and limiting the penetrating power and range of the shot. These benefits are minimized in close-range indoor situations, however, and some experts feel they are outweighed by the weapon's size and weight related disadvantages. A short-barreled 12-gauge shotgun, loaded with birdshot (not buckshot), is best for home defense. This is a powerful gun with a fair amount of recoil; it has a blast like an M-80 Firecracker, which in itself can cause damage indoors. If this bothers you, a 16-gauge gun is a bit milder but also lethal at close range, although ammunition is harder to obtain. A double-barreled or pump action gun is best, even though it will rarely be necessary to fire more than once, if at all. The mere sight of this cannon aimed at an intruder will cause his rapid surrender, unless he is irrational.

Two books on the subject of armed and unarmed self-defense are *In The Gravest Extreme* by Massad Ayoob and *Principles of Personal Defense* by Jeff Cooper. These and other works by experts on firearms and self-protection are highly informative.

Properly employed, a gun is usually not fired, but is used to nonviolently subdue a criminal and hold him until the police arrive. Never attempt to effect capture of a criminal unless you are completely confident that you can handle the situation and that there are no accomplices who can come to his rescue. Leave this dangerous task for the police, but if you must, here's how:

From a safe distance and from behind cover such as a door frame (if it's not bulletproof, it's not good cover), clearly and concisely order the perpetrator to "FREEZE." Say that one word, no more, firmly, but do not shout loudly. Make sure you are aiming for the heart, ready to fire, before speaking. Do not let your guard down because the intruder seems frightened or surrenders immediately; he may be seeking to distract you, and his gun may be close at hand. Do not try to frisk or even get close to him, especially if you are holding a long gun. Holding a perpetrator at gunpoint requires making it harshly clear to him that any attempt to fight or escape will result in instant death. You and only you are in command; he is to shut up and do as he is told. If possible, have him lie face down on the floor with his arms straight out over his head, or otherwise in a "spread-eagle" posture against a wall. Neither you nor anyone else should come within six feet of him, even if he is injured and appears helpless. Keep the gun aimed at his midsection and do not allow yourself to be distracted for even a split second. If he talks, demand silence. Upon the arrival of the police—and not before—put away your weapon. Show the police your gun permit if necessary, and clearly explain the circumstances of the situation. Be careful not to state anything that may be interpreted in a way that could be used against you and in favor of the criminal; make it clear that you were justified in the actions you took.

Should you ever shoot an intruder, you are likely to be sued by him should he survive or by others should he not. Never shoot at someone fleeing from you unless he stops to fire and you cannot take cover; it is not self-defense to shoot someone in the back. Never fire unless a life depends on it, but if you must shoot, shoot to kill. Should this be done, stay calm. Immediately call the police and your lawyer. Take the names of any witnesses who can speak in your behalf, and do not talk to strangers or the press, except on the advice of your attorney. You acted in self-defense so you need not fear the police, but choose your words carefully.

For information on gun dealers and firearms training check your Yellow Pages under "Guns" and "Rifle and Pistol Ranges." Choose a range that offers a course in basic firearms skills and will assist you in getting a license, if required. Be sure to investigate your legal responsibilities; they vary greatly depending on locality.

It would be irresponsible to discuss guns without giving some basic information on gun safety. When you understand how guns function and acquire experience in their safe handling and use, fears vanish and you have mastered a skill that can save your life and the lives of others. Lives are in jeopardy only when a gun is in the hands of the untrained or the criminally insane.

* Treat every firearm as if it were loaded, no matter how certain you are that it is not.
* Never point a gun at a person under any circumstances other than when necessary for self-defense, or to stop the person from committing a criminal act.
* As soon as you pick up a gun, make sure it is not loaded, unless it is to be used immediately.
* Keep guns unloaded when not in use and store the ammunition separately.
* Keep guns and ammunition out of the sight and reach of children and untrained adults.
* Weapons in storage or on display should be equipped with key-operated trigger locks, and the keys must be kept secure in another location.
* Never handle or permit anyone to handle a gun while intoxicated.

Guns and other weapons should never be displayed openly in a residence. Weapons are high on the burglar's list of preferred loot. A home that a professional might otherwise pass up will be expertly broken into if valuable guns are known to be kept there.

It is dangerous to keep a loaded gun ready for quick use against intruders if there is the least chance of the gun falling into the wrong hands. In most cases, there is insufficient need to justify the risk of a loaded gun. Whether or not you take the risk is a serious decision you must make based on your circumstances. You will usually have enough time to load the gun when needed, if you practice doing so quickly. If you do keep a weapon near your bed at night, always make sure to properly secure it in the morning; *DO NOT* leave it under the bed or in a drawer where it can be found by a child or a burglar. If you put down a loaded gun, make sure the safety is on to prevent accidental firing. As soon as you pick up a gun for possible use, set the safety to firing position. Be sure to return it to the safe position. Do all you can to prevent the need for a gun's use, but just in case, be prepared.

5

ALARMS!

The sophisticated criminal may know all about alarms and how to defeat the more vulnerable sort, but a good alarm system properly installed and used is about the best form of protection available at a reasonable price. You should never depend on alarms alone, but unless you have very little worth stealing, you shouldn't be without one. No thief will hang around long when bombarded with 130 decibels of noise from several inaccessible sources inside and outside the premises, lights coming on all around, and the possibility of police or armed guards showing up at any minute. The double-edged sword of technology has given us means of self-protection as well as self-destruction.

The story is told of the proud owner of a new Cadillac, equipped with a "state of the art" alarm system that would sound if anyone even sat on the car's fender, and could be heard a half-mile away. On the very day the car was driven home from the showroom the family's supper was interrupted by the sound of the alarm from the driveway. Gun in hand, the father ordered the wife and children upstairs, then went out to confront the would-be thief. After seeing no one in the area, he turned off the alarm with his key and closely inspected the car. He found no visible evidence of a break-in attempt, so he reset the alarm and went inside. At about 10 PM this performance was repeated, and again at 12:30 AM. Disgusted, frustrated, and convinced that the alarm was defective, he resolved to take the car back the next day for servicing, and left the alarm off for the night. In the morning the car was gone, stolen by the clever thief who triggered the alarm earlier. Good things can be too good.

A magazine on electronic design featured a nine-page article in a 1977 issue on a sophisticated microcomputerized home-security system. This one had it all—from a display panel in the master bedroom showing precisely where the intruder was, to smoke detector tie-ins, to button panels outside the doors requiring a five-digit code to enable entry without triggering the system. Pages were devoted to the software (programming) for the microprocessor. Surely this technological masterpiece, this Son of Star Trek, must be as close to foolproof as we have yet come! An affluent homeowner would gladly spend thousands of dollars for this marvel, and undoubtedly some have. Those who did got robbed even before the burglar came. The article itself warns that the biggest challenge of the project is finding the bugs that inevitably show up after assembly. Greater complexity generally means lower reliability, but in this case that's the least of it. No information is given on how to supply operating power to the system in a fail-safe way. Worst of all, simple open-circuit contacts (explained later) are used as intrusion sensors. These are the easiest of all to beat—by the mere cutting of a wire! Some microprocessor-based alarm control units for the home are currently on the market, hopefully with the bugs exterminated. The benefits, if any beyond fancy frills, are not evident in the specifications or price, and even if the greater complexity doesn't reduce reliability, it certainly does increase difficulty of quick repair for a do-it-yourselfer or a technician not trained in the new technology or lacking proper parts and equipment. Someday soon they will market a microcomputerized flashlight. If you rush out to buy one, please contact me as I know of a good deal on the ownership of a well-known bridge in Brooklyn.

WHAT TO LOOK FOR IN AN ALARM SYSTEM

What are the criteria for evaluating an alarm system?

1. Reliability. The system must always operate as it should, when it should, and never when it shouldn't.
2. Invulnerability. Efforts to defeat it should either have no effect or cause the system to be triggered.
3. Effectiveness. Performance certain to scare off the most brazen intruder and/or summon help quickly.

Just as you would not buy a home or car without taking pains to assure its value, you should be sure of the quality of any device used to protect them. In the case of the ex-Cadillac owner, the alarm was lacking in reliability, as a false alarm could easily be caused. The microcomputerized home system was vulnerable because of the open circuit sensors, the possibility of an inadequate power supply, and the push button code entry system (discussed later). Due to its complex design it may lack reliability as well. As no information is given on the signalling devices to use with it, it may also prove ineffective if these are not properly chosen and utilized.

THE THREE BASIC PARTS OF ALL SYSTEMS

Intrusion alarms have three basic parts. The first consists of the *sensors* that detect an unauthorized entry or entry attempt. Next is the *control,* the central electronic unit that interprets sensor signals and operates the alarm devices. Finally the *alarm signalling devices* themselves that produce noise and light, and/or summon help via a communications channel. "Silent" alarms only summon help, but residential alarms should make as much noise as possible. All three parts—sensors, control, and signalling devices—are subject to the quality criteria of reliability, invulnerability, and effectiveness.

No good alarm is cheap, although high cost doesn't prove high quality. Considerable savings—as much as 50 percent or more—may be realized by doing your own construction or installation. You should be somewhat experienced in wiring and general household repairs before attempting this though. Unfortunately, many systems sold for do-it-yourself installation are not as good as professionally installed ones; if you have the know-how to put in a professional quality system, you are at a great advantage. One important advantage of installing your own system is the added security of no one else knowing how the wiring runs or where the parts are located. There have been cases of installers who built weaknesses into a system and came back to burglarize the premises at a later date. If you hire an installer, check that the firm has been in business for several years, has good references, and a clean record with the police, Better Business Bureau, consumer protection agencies, the National Burglar and Fire Alarm Association, and any state or local alarm association. If your insurance requires an Underwriters Lab certified security system, be sure that the installer is qualifed to do such work and can issue a certificate of compliance. The alarm described in the next chapter is a "most for the money" unit for the build-it-yourselfer, and properly installed, is as good as many selling for several hundred dollars.

Unless you are a criminal, you know that you can't get something for nothing in this world, and sooner or later you'll be sorry you tried. If some particular security system or device is a "good buy," but is lacking in any way, be willing to spend more to get something better. The cost of an alarm system is a one-time insurance premium that aims at the prevention of a crime rather than compensation after the fact. It can save you from losing the fruits of years of labor; it can even save your life and the lives of your loved ones. What is that worth?

RUBE GOLDBERG ALARMS

There are hundreds of ways you can rig a contraption to make a loud noise when a door is opened. As the noise does not continue, it may startle an intruder but isn't likely to scare him away. Such gimmicks are crude and unreliable, and are, at best, a makeshift way to secure a hotel room door so that anyone entering uninvited alerts you. Unless you keep a weapon under your pillow when you travel, and can handle such a situation, you would be better off pretending you are asleep and not risking your life for a few dollars. Anything valuable should be kept in the hotel safe, and your wallet or purse should be hidden in your pillow, where it can't be stolen while you sleep.

Booby trap devices that can kill or injure an intruder (other than by frightening him to death) should NEVER be used. You may wind up killing your spouse, child, or a policeman responding to a burglary-in-progress call. If this happens, not only do you cause a tragedy, but you are more of a criminal than the intruder and you are likely to be prosecuted! Devices are on the market that squirt mace or tear gas at anyone who triggers them. Use of these in the home is unwise, as such things cannot distinguish between friend or foe. The burglar may deserve it, but would you want to inflict such pain on a family member? (Never mind your mother-in-law.) Would you want to be sued for macing a fireman responding to an alarm? These gadgets can get you in plenty of trouble. Avoid them.

LIVING ALARMS

Dogs make wonderful pets, but the typical pet is ineffective against intruders. A trained guard dog is expensive to buy and keep, and can be dangerous if not trained well, and there are many less than competent trainers in the business. If your dog bites the wrong person and you do not have good liability insurance, you may have to pay a costly settlement. Dogs, no matter how well trained, are stimulus-response animals without the reasoning ability necessary to evaluate a situation and distinguish friend from foe among strangers. People who rely on a dog usually do not get an alarm system (the dog may inadvertently trigger it) and this is just what some burglars want, as it is far easier to disable or kill a dog than to bypass a good alarm. And will your living alarm care for itself when you go on vacation? An electronic security system never has to be fed or taken for a walk, it won't defecate on the sidewalk, and if your alarm ever has puppies you'll be witness to a genuine miracle.

Dogs are not recommended for residential security use, but if you want one for a pet anyway, you might as well consider its limited protection value and carefully select one accordingly.

The alarm dog, most suitable as a pet, is one that need not be large but must have a loud bark. This type does not bite but only makes noise; it does not require extensive training and will deter some amateur burglars. To be most effective it should be trained to make noise only when necessary, and like all security dogs, it must have free access to all parts of the protected home or area. Even though its bark is worse than its bite, let everyone but your immediate family believe it's vicious. To build a mean reputation you might even muzzle it when you take it for a walk.

The guard dog provides security with teeth, but is not an ideal pet. This is a large, costly dog, usually a German Shepherd or Doberman Pinscher. It must be of good breeding and thoroughly trained by both its owner and a qualifed professional, and the owner must be adequately taught how to control it and train it further. These dogs deter the average burglar, but are a liability to the owner. Adequate barriers must be present around the protected area to prevent escape and possible attack on an innocent person.

The attack or patrol dog is a vicious killer and must *never* be kept in the home.

If you want a guard dog, you should get a young one and have it competently trained. Police departments with canine divisions and the American Kennel Club can refer you to a qualifed trainer; check the trainer's references, experience, and certifications. Personally inspect the training facilities to see that the dogs are healthy and clean, and are kept humanely. Check the company's record with the Better Business Bureau and consumer affairs agencies. Make sure to get a written contract giving the specific service, fee, and terms of liability.

SYSTEM TYPES AND FEATURES, GOOD AND BAD

Alarms are of two basic types; *perimeter* systems that are operated by sensors on doors, windows, and other entry points, and *interior* or *motion detector* systems that respond to the presence of an intruder in the protected space. Perimeter systems may be either "hard-wired" (sensors connected directly to alarm control circuitry) or wireless (remote), where sensors transmit a radio or other signal that is detected by the central control. Motion detector systems take many forms—microwave, infrared, ultrasonic, photoelectric, sound sensing, and proximity (this last kind is not strictly defined as a motion detector). All of these have their uses, but the best system for residential protection is, by far, the expertly installed hard-wired perimeter system. It is the most reliable for the money, and unless badly designed, installed, or inadequately equipped with sensors, it is the least vulnerable. With a good array of alarm signalling devices, it is more reliable and effective than other systems similary equipped because it is far less prone to false alarms. Systems that cause false alarms are practically useless; they are ignored when the alarm is for real and their owners tend to leave them unused. No amount of repair can correct this if the system type is not suitable for the application. The perimeter system can protect you when you are at home sleeping, taking a shower (remember the shower seen in the movie "Psycho"?), brewing moonshine in the basement, or anything else. Some criminals *will* enter your house or apartment *while you are home* with the intention of taking what they can and escaping undetected. They can be dangerous if accidentally confronted, but a good perimeter alarm in use can prevent their entry. The interior system can only be used when no people or pets are at home, as normal activity will trigger it. The ultimate in protection is a good perimeter system backed up by reliable interior sensors that can be switched off easily when the home is in use.

Many motion detector alarms are small, self-contained units that are portable and easy to install. This is the one advantage they have over perimeter systems, but is also partially responsible for their greater vulnerability and lack of reliability and effectiveness.

Let's take a look at a few basics of electrical connection necessary to the understanding of what follows.

All sensors, whether activated by a magnet, mechanical button, or a sophisticated electronic sensing device, work by either closing a contact so that an electrical current can flow through it, or by opening a contact, thereby interrupting a current that was already flowing. Every such contact has two terminals, that we will call "A" and "B" here for descriptive purposes. If we connect all the A terminals together and all the B terminals together, the contacts are said to be connected in *parallel.* If any one of them closes, current can flow through it between points A and B. Sensors are wired this way in an *open circuit* system, meaning no current flows from A to B until a contact closes, and when this happens, the alarm will be triggered. If, instead of this arrangement, we connect the B terminal of one sensor to the A terminal of the next, we form a chain of contacts called a *series* connection. If all the contacts are closed, a current can flow through the chain, but if any one of them opens, the current is stopped. Sensors are wired this way in a *closed circuit* system, which triggers the alarm when the current path is broken. Figure 5-4 shows these connections.

Now let's examine the features that are available in alarm systems and their relative value:

Very Desirable

1. A backup battery that can power the system for several days if normal power fails or is cut off. This battery should be rechargeable and be kept fully charged by the alarm circuitry when normal 115V power is available.

2. Tamper switches. These cause the alarm to sound if an attempt is made to defeat it by either opening or removing the control unit, keyswitch, bell, siren, horn, or other parts. The switches should work even if the alarm is not on when the tampering occurs.

3. Provision for connecting and ability to operate many external alarm sound and light devices. Although more is desirable, 2 amps at 12V DC drive capability with "dry" relay contacts rated for 5 amps at 115V AC is usually adequate.

4. A keyswitch system on-off control with a high-security pick-resistant cylinder and tamper switch. If the keyswitch is outside the premises, its protection is especially important; mounting it high on the door or door frame helps to discourage picking attempts. If the switch is of the momentary closure type, a concealed second switch in series with it gives added security (as described in Chapter 2). With an ordinary on-off switch whose contacts remain closed while the alarm is on, a parallel connected backup switch is advisable (as described in Chapter 6). A digital keypad control operated by entering a code is superior to a keyswitch ONLY if it is constructed and operated so that it is impossible for another person to observe the code.

5. Entry/exit delays. These allow time to leave the premises before the alarm is armed and time to shut off the alarm before it sounds when you enter. When present, an alarm on-off control that is operable from outside the premises is unnecessary.

6. Automatic alerting of a central alarm station or the police by means of a secure phone line or radio transmission. This is essential if your home is in an isolated area where an audible alarm may go unheard.

7. Sensitivity control on motion detector systems.

8. Battery low indicator if unit is powered by battery alone. (Unless for short-term outdoor use or a "wireless" remote transmitter, no alarm should be powered by battery alone.)

9. Comprehensive instructions written by the manufacturer on installation, operation, and maintenance.

10. Written warrantee from the manufacturer or installer with clearly defined terms.

Desirable

1. Panic button or switch. Usually located by your bed, this immediately sounds the alarm when activated. The sensing circuit or "zone" to which this is connected is active twenty-four-hours a day and may also be used for tamper switches. In a large home, several panic buttons are advisable throughout the premises.

2. Automatic alarm cutoff and rearming of the system after sounding for at least five minutes. This feature is not as desirable for closed circuit perimeter systems without backup protection, as will be explained in the discussion of sensors. For such systems, a backup "loop," or a zone that operates only when necessary, is a desirable feature.

3. Provision for connecting open and closed circuit hard-wired sensors even if the system is not the perimeter type.

4. Sensor monitor circuitry that tells if a window has been left open or some other fault exists in the sensing loops before the alarm is activated. Some systems have a green status light that indicates that all is well.

5. Delay override switch or "instant" mode. This causes the alarm to sound immediately when you are at home, the switch is on, and a door is opened that would normally cause a delayed alarm.

6. An internal power supply from which 12V DC is available at all times to power sensing devices such as passive infrared that may require such power. This should be a regulated (constant voltage) supply, preferably with a current capacity of over 300 ma. (0.3 amps).

7. Separate system zones for large homes or apartments. One zone can be used for perimeter protection and another for backup, or different areas may be on separate zones. Such systems should have an indicator panel that shows which zone has been triggered.

8. Alarm memory indicator. If on, this tells you that the alarm was triggered and has shut off and rearmed itself. This indicator should remain on until manually reset, even if the system has been turned off.

9. Provision for connecting remote alarm memory indicators. These should be near all outside doors, so that you can tell before you enter if the alarm was triggered. If so, an intruder may still be present and you should not enter, but call the police.

10. Provision for smoke/fire detector connection with an alarm signal that is easily distinguishable from an intrusion-caused alarm.

11. LED system status indicators inside the premises.

12. Lightning/high-voltage circuit protection.

13. Fail-safe arming that prevents the system from being activated in the event of a sensing-loop fault. This makes it necessary to either correct the fault or bypass the zone it is in before the system can be used. "Faults" are often no more than a window left open, which can be remedied immediately. If a sensor or wiring defect exists it should be repaired immediately, but if this cannot be done, switching out that one zone allows the rest of the system to function until it can be. Part of the system working is better than none at all, and at such times interior backup sensors on a separate zone are invaluable. Fail-safe arming may not be desirable in a simple unzoned or two-zone system, where a single fault can render most or all of the system unusable. Here a fault-indicating LED is sufficient.

Undesirable

1. A noisemaking device contained in the control box with the alarm circuitry. This immediately leads the brazen intruder to the heart of the system where he may be able to defeat it. An alarm with such a self-contained noisemaker should have provision for its disconnection and the use of external signalling devices.

2. A digital keypad (push buttons) for entering a code to arm or disarm the system from outside the premises is not a good feature unless it is constructed and used so that the code cannot be observed by anyone else, whether standing beside the operator or at a distance using binoculars. A view shield must surround the keypad, and while operating it, you should stand close to block the front with your body. The code should be easily changeable and be changed periodically; it

should be changed immediately when there is a chance that it has become known to someone who has no business knowing it. This type of control inside the home in a concealed location is very good; probably more secure than a keyswitch control.

3. Remote alarm system on-off (arm-disarm) controls. These are convenient but add vulnerability, especially when located outside the premises. If the system has entry/exit delays, it is best to have only one control that is part of the central control unit or panel mounted in a concealed location. In a large home where a single on-off control may be impractical, put the remotes in hidden spots and/or disguise them. Further details here would tip off the enemy, so use your imagination.

4. Indicators outside the premises (other than a remote alarm memory indicator) that tell the status of the system in a way that a stranger can recognize, such as red for on and green or white for off.

5. Entry delay warning device. This tells an intruder, when he comes through the door, that an alarm will sound in a half minute or so. If it scares him away it's fine, but some professionals will just take it as advance warning and try to disable the system before the main alarm sounds. If you feel you must have this feature to remind you to shut off the system before it sounds, locate the prealarm sounder far away from any other part of the system. When used in this way, it helps prevent false alarms.

6. A "shunt switch" that takes a door out of the system's coverage by disabling its sensing switch. It may be good while working in the yard or garage to have the rest of the house protected, but it is easy to forget about with the result that one entrance is left unprotected later. If such a switch is used, it should turn on an indicator to remind you to reset it when no longer needed.

7. Alarm warning stickers on doors and windows. These may give the name of the manufacturer or installer and may enable the perpetrator to learn a way of defeating the system. The element of surprise, important in scaring off an intruder, is also lost. Stickers deter some burglars, but not as much as alarm tape on all the windows. They are often merely a bluff, and burglars know it.

Before getting into specifics on the more desirable alarms, let us take a look at why some others are less than satisfactory for residential protection.

Ultrasonic

The story is told of Neb, the condominium owner, who had owned various cheap alarms only to discard each when it malfunctioned, usually by causing false alarms. Finally he heard a crass radio ad for Insane Irving, who was having a sale on ultrasonic alarms. NOW! PROTECT YOUR HOME WITH THE NEW INVISIBLE SHIELD! TECHNOLOGICAL BREAKTHROUGH! DON'T WAIT TILL IT'S TOO LATE! GET YOUR SYSTEM NOW! IT'S INSANE! Not the sort to know when he is being manipulated, Neb went to Insane Irv's local "funny farm" where the salesman, one Mr. Slicke, removed all his remaining doubts and vestiges of logical thinking. For a mere $299.95 (a nickel makes a big difference to Neb), he was the lucky owner of the device that was to put an end forever to the housebreaking profession. He put his new toy on a shelf near his door as directed and plugged it in. A shrill alarm sounded after the thirty second exit delay as he moved in front of the unit. He spent the next hour playing with the sensitivity control, making sure the unit would work but not cause false alarms. Overjoyed, the next day he boasted to co-workers of how the new alarm brought him peace of mind. Lloyd seemed particularly interested, saying he had been burglarized and wanted one like it, so Neb gladly answered all his questions. The next Saturday night, Neb lost about $3500 to a burglar. Neb trusted Mr. Slicke, but not banks. Lloyd quit his job a week later, able to get along quite well without it for awhile. His off-the-books work was more profitable anyway, and as long as people used alarms that could be beat by moving very slowly until the plug was pulled, he would have work whenever he wanted it.

Ultrasonic systems for the home have their defenders, but while fine for some applications, they are not optimum for most residences. These work by emitting inaudible sound and listening for changes in the sound field caused by a moving body. Properly utilized they are hard to defeat, but they are very prone to false alarms due to the noise from a ringing phone, doorbell, or from outside the premises, air motion, normal vibrations of walls, windows, floors or ceilings, and from temperature and humidity changes. They may not work at all in acoustically dead areas. Ultrasonic sensors of more sophisticated design have reduced some of these problems but require installation by a professional for greatest reliability. Readjustment is also required whenever a change is made that affects room acoustics, such as the installation of rugs, drapes, or the movement of furniture.

Sound Sensing

Many of these are little more than a fraud on the consumer. Some are heavily advertised and claim to be able to hear a burglar's noises and scare him off before he even gets in. They are triggered by high-frequency sounds that can come from many sources other than a break-in attempt, such as car brakes, fire truck sirens, your doorbell or phone, or a neighbor's pride and joy ultrasonic burglar alarm. Should the latter occur, the noise of your alarm may then trigger his, and they will blissfully continue to turn each other on ad infinitum. False alarms abound, and self-contained units are easy to defeat as well. If you buy one and decide you want your money back, chances are the company that marketed it will be gone as fast as a thief. Caveat emptor!

There are some higher-quality sound detectors made and marketed by reputable firms that can be used in vaults and *quiet* storage areas, basements, garages, and similar places. In general, however, sound sensors are not the best choice for an apartment or in most areas of a home.

One place that sound sensors are useful is in the automobile, where even a rank amateur knows you can bypass the standard siren alarm by breaking in through a window. Unless the car is parked in a quiet location, such as an indoor garage, false alarms from time to time are inevitable. The alarm should be operated by a hidden switch in the passenger compartment and have entry/exit delays and automatic shutoff/reset. It should operate independently from the siren alarm, preferably by sounding the horn in an on-off-on-off signal. One drawback is that the system draws a small but constant current from the battery, and may run it down if the car is not driven for several weeks.

Also valuable for auto protection are systems that employ motion or tilt sensors that trigger the alarm if an attempt is made to jack up the car or tow it away. They can cause false alarms, however, if excessively sensitive. Hidden ignition and fuel cutoff switches inside the car are often more effective than expensive commercial anti-theft devices. Make sure to run all security system wiring where it cannot be seen or reached from outside or under the car, and use an extra hood lock such as the Chapman lock to hinder defeat by battery disconnection. If you want a siren alarm, a warbling electronic siren is more efficient and effective than a motor-driven one. This draws only about 1/2 ampere of current, as compared to the well over 10 amps of a motor siren.

Photoelectric (Visible Beam)

They are unreliable, false-alarm prone, and defeatable with a flashlight, reflector, or by avoiding the beam. For security applications they are practically obsolete.

Doorknob Alarms, Alarmed Chain Locks, and So Forth

These cheap toys are easily defeated, and many are false-alarm prone. All the intruder need do with a hang-on unit is smother it under a pillow and it will be inaudible outside the room. The better devices of this sort are useful for awakening you should someone enter your hotel room, but are a poor choice for home security applications.

Microwave ("Radar")

Like the ultrasonic alarm, this operates by setting up a weak energy field in the protected space and sensing when a moving body alters it. Microwave sensors when professionally installed are versatile for commercial, industrial, and even outdoor applications, but are less desirable inside the residence. Unfortunately microwaves can penetrate glass, plastic, and thin walls so that false alarms may be caused by motion outside the bounded area. Electromagnetic radiations, such as those of fluorescent lights, can also cause false alarms. If plastic pipes are used, an upstairs neighbor can trigger your alarm by flushing the toilet. Enough said.

Pulsed Infrared (IR)

This is a photoelectric system using a modulated (signal carrying) invisible beam. This is better than visible light systems but subject to some of the same drawbacks, such as false alarms due to the beam aim being upset by normal vibrations. If used, the transmitter (beam source), receiver, and/or reflectors must be very securely mounted to rigid supports and carefully aligned optically. They must never be exposed to direct sunlight. Dust accumulation on the optical surfaces should be removed from time to time, and alignment should be checked after this is done.

Closed Circuit TV (CCTV)

CCTV is excellent for banks, large stores, and high security apartment complexes, but usually too expensive and impractical for single family residences.

To sum up, all of the above have their uses, but are generally not best for residential protection. Unfortunately, many are sold for home use, and are promoted with no mention of their flaws. You must insist on what is best for you, and not be deceived.

Now we get to some system types worth considering for home and apartment.

Passive Infrared

Probably the best of the motion detecting systems for residential use, passive infrared can be used as a backup for a perimeter system. Of course it cannot be used when the home is occupied, so it must be controlled separately from the main system. Passive infrared detectors "see" both the body heat of an intruder and his motion, and trigger only when both are present. The area over which the detector senses can be large and should be adjustable by means of its lens assembly. This is not as prone to the false alarms that plague other motion detectors, but care in adjustment and choice of its location is required. A passive IR unit should have a walk-test LED indicator that allows you to adjust the detection field by actually seeing when it is triggered as you walk through the space it covers. Its location and coverage pattern should be such that an intruder will be moving across its field of view, and it must avoid windows, stoves, radiators, fireplaces, hot air ducts, heaters, and other possible sources of infrared energy. A desirable feature in these sensors is an environmental test switch that boosts its sensitivity so you can test for possible sources of false alarms.

Proximity

This is an electronic device that triggers the alarm when the metal object it is connected to is touched or even closely approached. This is good protection for safes, filing cabinets and metal desks that can be insulated from the ground, this being necessary for the system to work. It can even be used on a car in a garage, as the tires insulate it from the ground. Be sure to disconnect it before driving away, or your alarm will pretend it is a watchdog and chase your car down the block.

Remote Sensor ("Wireless") Perimeter Systems

These systems use sensors at doors, windows, and other points that transmit a signal to the control receiver, which sounds the alarm. Reliability depends on design, installation, and maintenance and is OK but is often not as good as a comparable hard-wired system. Wireless systems are easier to install, but more expensive than a wired system. They may be used with sensors in a garage, tool shed, or other structure up to a few hundred feet away without running wires. Be aware though that the sensors will become useless if the batteries are not replaced regularly. Possible battery failure is the biggest reliability problem of wireless alarms. A system using coded signals may be zoned to indicate which transmitter caused the alarm, and such coding reduces the possibility of false alarms or defeat by a sophisticated burglar with good electronic equipment. Portable pocket panic button transmitters are available for wireless systems, an excellent feature for high-crime areas where one might be attacked upon entering or leaving home.

Rather than using radio transmissions, the sensors of a few "wireless" systems *are* wired—not directly to the central control, but to the 115V power wiring. The central control is wired likewise, in order to pick up the sensor signals as well as obtain operating power. The power lines carry the signals, and, in effect, make the installation of "hard wiring" unnecessary. Such systems are often unusable, as there may be several separate power lines running into your home or apartment, and sensor signals on one line may not be able to get through to the control unit on another. It is also possible, especially in apartment buildings, for your equipment to interact with someone else's. If you buy a system of this type, be sure to get a written guarantee of a refund if you must return it.

Wireless systems are easy to install yourself, and should come with clearly written instructions. Keep in mind that the central control unit contains a radio receiver that responds to transmitters operated by the sensors. Both control and transmitters must be placed away from large metal objects and never inside a metal enclosure, such as a tool cabinet or metal shed. The transmitters must be accessible for weekly battery tests, and should have test buttons and indicators that permit easy testing without triggering the alarm. Before wiring the noisemaking devices to the control unit (the sensors may be wireless but these are not), thoroughly test the system's response to every sensor. As the transmitters are fairly expensive, you may wish to wire all the sensors in one room to one of them, rather than having one transmitter at every window and other sensing point. Such a "semi-wireless" system is a compromise between ease of installation and cost. Higher cost, the inconvenience of frequent battery testing, and inherently lower reliability due to possible battery failure make hard-wired systems preferable to wireless ones, in general. Such systems are becoming better with technological advancement however, and some new ones may rival hard-wired systems.

Hard-Wired Perimeter Systems

A hard-wired perimeter system is the most reliable and invulnerable home security system one can get for the money. On the rare occasions that such a system is defeated or ineffective, the cause is usually inadequate or defective sensors or alarm signalling devices, or improper installation. Perhaps the only drawback of these systems is the installation difficulty. Unlike motion sensing systems however, it is easy to expand them at minimal cost by adding more sensors at a later date. If you are planning on building a home or any other structure requiring security, the best time to install wiring and sensors is during construction when they can be most effectively concealed. Security should be a prime consideration in the architecture of new structures, and a wired-in alarm system is highly cost effective. A good feature to build into your power wiring is a master switch that turns on all the lights in the house at once. This switch should be located by your bed, and also be operated by a relay connected to your alarm system. Outdoor electrical outlets are *not* desirable, especially if located near a door or window. You wouldn't leave out a power tool for a burglar to use, so don't leave out the power for a tool he may bring. You might consider a booby trap outlet though; this is not connected to your power lines, but is wired to your alarm so that it is triggered by anyone trying to use the outlet.

Sensors for wired systems are either open circuit or closed circuit. Open circuit sensors are all wired in parallel. When any one of them closes its contacts, the circuit is completed and the alarm is triggered. Unfortunately, they can be defeated easily just by cutting a wire. Closed circuit sensors are wired in series, and the protective loop of sensors carries a small current. If any one is activated, this circuit is broken and the alarm is triggered. If a wire is cut in an attempt to disable the system the alarm will sound, but these sensors can be defeated by short circuiting. This is more reliable than the open circuit system, as any break in the sensor loop is immediately apparent. Some alarms feature sensor monitor circuitry and a test mode to indicate such a problem. A break in an open circuit system can only be found by testing each sensor individually. Such a defect—accidental or deliberately caused—can go undetected until it is too late. For residences the closed circuit system is most commonly used, and is quite adequate if wiring and sensors are concealed and inaccessible from outside the protected premises. Most alarm control units have provision for the connection of both open and closed circuit sensing devices.

An "end-of-line (EOL) resistor" loop found on some alarms is a sensor circuit that can accommodate both open and closed circuit devices. A resistor at the end of the two-wire line permits a small "supervising" current to flow. Open circuit sensors when operated short out this resistor, and the resulting higher current triggers the alarm. Closed circuit sensors are wired in series, and when operated trigger the alarm by breaking the loop. The overall circuit is quite secure, but individual sensors may still be defeated by shorting or wire cutting.

The old terms "normally open" (N.O.) and "normally closed" (N.C.) used in reference to sensor contacts are fortunately rarely found now, as they have caused endless confusion. "Normally" refers to the unactivated condition; with a button switch, this means when the button is *not* held in, as when the door it is installed in is *not* closed. Thus, the contacts of a "normally open" switch are closed when the door is closed, and it is wired into a closed circuit loop of sensors in series. When the door is opened, the circuit is broken, triggering the alarm.

What was called "normally open" is now called "closed circuit," as that is the type of system such sensors are used in; likewise "normally closed" sensors are now usually known as "open circuit" and are for use in open circuit systems. Remember them by the opposites rule, normally open for closed circuit and normally closed for open circuit. Even this is not universally accepted, however.

SENSING DEVICES

The previously mentioned motion detector types of alarms are generally available as small self-contained products with sensors, control, and noisemaker in one box, and also with sensors only, to be wired into a perimeter system for interior backup. The self-contained units of all types are generally not worth the money, no matter how cheap they may be. High-quality sensors of the passive IR and proximity type are worth considering.

Unlike the switch contact type of sensors discussed later, motion detectors, such as passive IR used for interior backup, require a supply of power, usually 12V DC. They either have their own transformer and backup battery or get their power through a connection to the supply in the central control. When these sensors operate, they open or close an internal relay contact that is wired into one of the control's sensing loops.

The variety of sensors available for use with open and closed circuit systems is vast, as can be seen by glancing through the pages of any security equipment and alarm supplies catalog. As a chain is only as strong as its weakest link, an alarm is only as reliable as its worst sensing device, so caution is necessary in their selection and use.

Recommended are magnetic switches, particularly the sealed reed type, which are very reliable. These can be used on doors and windows, behind picture frames, behind drawers--their applications are limited only by your imagination. They are installed so that a magnet is normally held near the switch, the magnetic field holding the contact open or closed. When the magnet is moved away by the opening of the door, window, or drawer, the switch closes if it was open or vice versa, and the alarm is triggered. To avoid defeat, they must always be out of view of a potential thief. Super high security magnetic switches are available that trigger the alarm if an attempt is made to defeat them with external magnetic devices. Small mechanical switches are useful if strong, well-concealed, properly installed, and weatherproofed (if they may be exposed to the elements). They are typically installed in door frames so that the door holds the switch button in when closed, and releases it when the door is opened. Heed the manufacturer's or retailer's installation instructions. Some of these switches have three terminals for connection, "COM.," "N.O.," and "N.C." Apply the opposites rule and connect to "COM." and "N.O." for a closed circuit system and to "COM." and "N.C." for an open circuit one.

Foil tape for use on glass is essential, as glass cutting, breakage, or removal is an easy means of entry. Glass breakage sensors may be used but are prone to false alarms from such things as a hailstone or pellet striking the window. For good protection you need one sensor on every pane, which is costly. If it is properly installed, tape is preferable for residential use. This tape is a lead/tin alloy and usually comes in a roll, gummed on one side. Aluminum foil is not recommended as it cannot be soldered and requires special connectors.

Tape installation instructions often are available from the retailer and may be found in some security equipment catalogs. Some people have trouble, but with a little practice with cheap masking tape you can learn to do it very neatly. You can have it professionally installed, but if so make sure they do it according to your instructions, as guided by Figure 5-1 and what follows here.

The tape must be applied to a clean, dry surface, must be subject to a SLIGHT tension to stretch it as it is applied, and must be pressed down tightly for adequate adhesion. There should be no more than six inches of glass between adjacent tape lines, and no more than six inches between the window sash and the nearest tape. It is a mistake to merely run the tape around the edge of a large pane, as this leaves a large central area unprotected. The best protection of all for large windowpanes or plate glass is to use *both* tape and a glass breakage detector.

Connectors that stick to the glass and wires that run to the pane are vulnerable; it is best to run the tape onto the sash (carefully insulated from it if it is metal), secure it tightly to the sash, and make the connections there. The two ends of the tape should not be close to each other where they connect to the wires, but should be on opposite ends of the pane.

A small low-power soldering iron may be used to connect wire leads to foil tape or to repair a break by soldering a thin wire over it. Excessive heat will melt the tape, but if the surface is clean and the iron temperature is not too high connections and repairs can be made quickly. If the tape surface is dirty or oxidized clean it with fine steel wool, and make sure to tin the wire with solder before applying it to the tape. Only rosin core solder should be used for all electrical connections.

If foil tape without gum on one side is used, it must be glued to the glass with lacquer. In this case, be sure you have all necessary materials and understand the manufacturer's instructions before beginning. Varnish for covering foil tape should be used if the tape will be exposed to rubbing or scratching.

Any window that is not necessary for ventilation or fire escape should be permanently sealed. In this case, after taping the glass, run the tape from the sashes to the wall or immovable frame in several places, and securely hold it there with glue or strong mylar tape. This protects against forced opening as well as glass cutting, without the cost of additional switches. The presence of foil tape on *all* your exterior glass is a good deterrent to most burglars. Figure 5-1 illustrates its proper application.

Shock, vibration, and stress/strain sensors may be used to sense break-in attempts through walls and ceilings, but are more susceptible to false alarms and are more costly than foil tape or thin wires. Tape or "lacing wire," for use in a closed circuit system as with window tape, can be run along vulnerable surfaces and then concealed with paint or other covering. Shock and vibration sensors have the advantage, however, of being able to trigger an alarm before much structural damage is done. Surface protection is recommended for all perimeter wall surfaces including those between garage and house, and for the roof or attic flooring.

Many devices are available for use in closed circuit systems to protect such possible points of entry as skylights, air conditioner ducts, basement windows or chutes, fences, chimneys, and so on. It is doubtful that any of these devices are better than a simple arrangement of foil tape or thin wire that breaks upon intrusion and sounds the alarm. A professional burglar may be familiar with devices on the

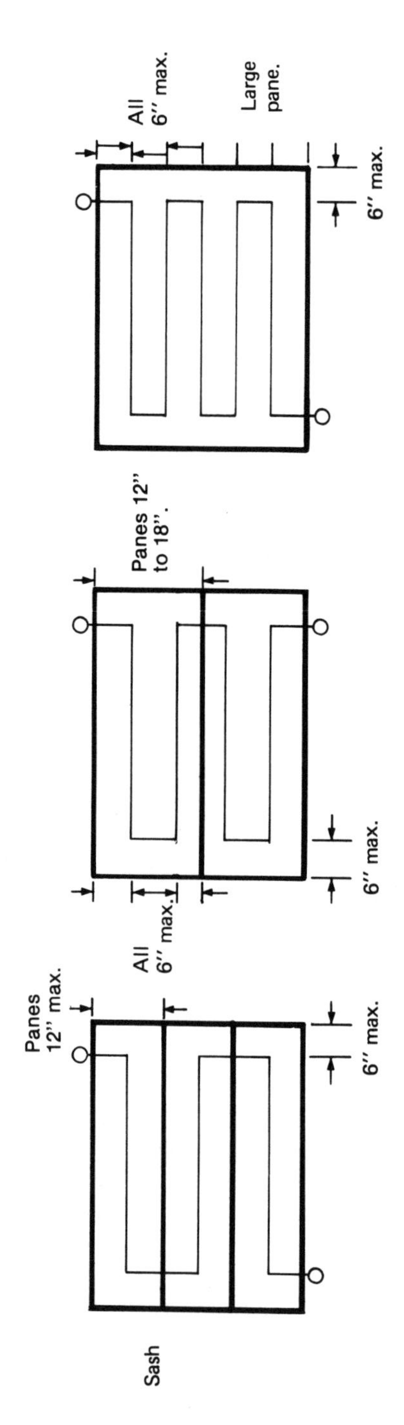

Figure 5-1. Window tape application. If sash is metal, tape must be insulated from it at all crossing and connection points.

market and have a plan for beating them, but he has no way of knowing the nature of a custom-made device. An easily broken wire or foil tape strip can be glued to a wooden slat that is then mounted securely across the opening to be protected, so that the wire or tape cannot be seen by an intruder. Breaking or removing the slat sets off the alarm. Properly made, this is more tamperproof than standard "pull trap" devices, and costs mere pennies.

Clever criminals may know how to bypass some alarm sensors, but they cannot avoid walking on the floor at least some of the time. One of the simplest and best forms of interior protection is by means of tapeswitches and pressure mats that are concealed under carpeting, in places an intruder is likely to step on and cannot circumvent. These devices are available in several forms for outdoor as well as indoor use, the sealed outdoor ones being quite rugged, and the indoor ones inexpensive and very reliable compared to many other types of interior sensors. As with all interior sensors used with a perimeter system, they must be disconnected when the home is occupied and the system is in use, and connected to the system when no one is home. Installation instructions must be followed carefully and care must be taken not to damage the contacting strips by bending. The pressure sensitivity of the switches can be increased by putting a layer of thick tape over each contacting strip; to decrease sensitivity, place the tape on both sides, but not over the strips. Materials may be purchased that decrease sensitivity enough so that a pet stepping on the mat will not cause a false alarm. Beware of welcome mat sensors as they are easily recognized and avoided by an intruder; to be effective the sensors must be invisible (see Figure 5-2).

Concealed switches on bedroom, bathroom, basement, and closet doors are good interior protection, but of course the doors must remain closed when this part of the system is in use. If you have a home with an attached garage, it is imperative that the door between garage and home be a very secure one, and it must be connected to the alarm with a sensor switch located on the side inside the home.

Outer garage doors should be equipped with alarm sensors but require a means by which you—and only you—can open the door without either triggering or turning off the system. Manually operated doors may be provided with a key operated shunt switch, but care must be taken to make this as secure as possible and to never leave the switch on so that the sensor remains disabled. If you have a remote control door opening system that can turn on the garage light when activated, you can connect a relay to the light operating circuit that will bypass the door sensor like a shunt switch when the door is opened with the control.

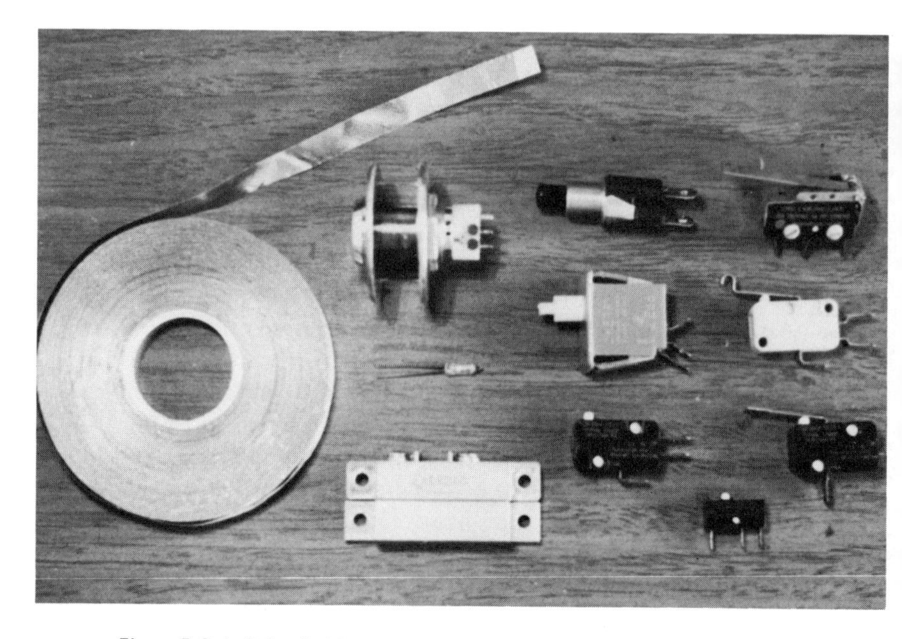

Figure 5-2. *Left:* A roll of foil tape. *Next left* top to bottom: Alarm key switch for a door, mercury switch tilt sensor, magnetic switch with magnet. *Right:* Mechanically operated switches.

Where closed circuit perimeter protection is used, it is desirable to have backup interior protection for maximum security. If, for instance, a skylight trapwire is broken, the alarm will sound and the burglar may flee, only to return some time after the alarm shuts off and resets. The closed circuit loop remains broken and is useless, so unless the premises is under guard, he may now enter. Motion detectors, open circuit sensors, or closed circuit sensors on a separate alarm zone can provide continued protection. Some control units have a "phase II" backup loop feature for this purpose. The beauty of this feature is that the backup sensors are only active when necessary and therefore cannot cause a false alarm when the home is in normal use. On the negative side though, should an intruder succeed in getting by the perimeter protection by other means, the backup sensors will not function if they are only connected to the phase II loop.

Interior sensors covering the path between the door and the system on-off control (if inside the premises) must be wired to the entry/exit delay loop, as is the door sensor, so that when you enter you can disarm the system before it sounds. All other backup sensors should go to the phase II loop, or preferably to an instant response (no delay) loop with a hidden switch to disconnect or bypass them when the home is in use and perimeter protection alone is desired.

Mercury switches, that close or open a circuit when tilted or shaken, can be used to wire theft prone appliances or large valuables into a security system. By using the panic switch circuit they can sound the alarm even when it is not on. False alarms, however, can result from any motion or vibration.

One means by which both perimeter and interior backup protection may be defeated is the shutting off of the entire system. This may be possible when an on-off control is located outside the premises or is inside, but is poorly secured. There are clever ways around even this vulnerability however, as well as inexpensive ways of adding interior backup to a perimeter system. Rapine discovered this upon returning to the *Gulls'* home several months later when they were out for the evening. She had met her new boyfriend Rick the Pick through her probation officer; indeed they had a great deal in common. Rick was a graduate of a mail-order course in locksmithing which had supplied him with the tools and knowledge he now generously bestowed upon his beloved Rapine in exchange for 25% of her swag. Tonight she worked alone, confident she was not being two-timed as Rick was also busy, hitting a luxury apartment building across town.

Archie had recently read a book titled *Prevent Burglary,* but upon adding up the cost of all the new doors, windows, locks, and additional alarm equipment necessary to make his home highly secure, he decided he would be smart and get the most for the least. Hours were spent with wirecutters and screwdrivers, furniture was moved and replaced, and the house reeked of soldering flux. Tiffy stared at him as the TV and easy chair were abandoned in favor of running wire under rugs and into holes made in chests of drawers, bookcases, kitchen cabinets, closets, the laundry hamper, and even the medicine chest. Archie stared back, increasing her fear for his sanity with such cryptic ejaculations as "Let 'em get in, they won't stay long!" and "We'll see who gets who!" Finally, his work completed, Archie took Tiffy out to dinner, coincidentally on the very night Rapine had planned her visit.

By patient observation Rapine had determined that Archie's nosy neighbors were out for the evening as well, so if anyone did see her at the door it would not be likely to be anyone that would know her as a stranger. Besides, women were above suspicion. Her practiced fingers guided the tools well; the cheap alarm keyswitch in the door took less than a minute. The mortise lockset took two, and there was no second lock. She worked in semi-darkness, with an occasional glance over her shoulder to be sure that she wasn't being observed. The door opened before her; she entered confident of emerging shortly with a small fortune.

She emerged far sooner than expected, moving as if physically propelled by the blaring din of the alarm she was so sure she had beaten. Away she ran to the comforting arms of Rick the Pick, whose boasts of success were interspersed with innuendoes about her lack of skill. Under his tutelage, he assured her, she would rise above such defeat. For now at least Archie's home was safe.

Archie's backup protection had cost him under fifty dollars. He installed a closed circuit loop of concealed sensors in many of the typical places a burglar searches for loot, as given in Chapter 7. He even arranged switches under his TV, stereo, and other appliances that would break the circuit when the item was lifted. These were connected to the coil of a sensitive (low current drain) 12V relay, and the loop was powered by the 12V DC supply that is on 24 hours a day in his alarm control unit. If a drawer is opened or the loop is broken by any other means, the current to the relay is interrupted and it closes a contact connected to the tamper switch/panic button circuit, which sounds the alarm whether the door keyswitch is on or not. Of course this system cannot be allowed to work when the home is in use, so Archie wired a shunt switch across the sensors to bypass them. Before leaving home, he carefully checks that all the drawers, doors, etc. are closed to prevent a false alarm, then he throws the hidden switch to enable the circuit to operate; when he returns home he cuts it off with the same switch. Figure 5-3 shows the circuit. As soon as Archie saw this, he knew it was just for him.

Here's how it works: The relay coil in Figure 5-3 is normally energized at all times, holding the contact between the "COM." and "N.C." contacts open. If the shunt switch is in the ON position, the opening of a sensor contact interrupts the flow of current through the relay coil from the always-on power supply in the alarm control unit. This allows the contact between "COM." and "N.C." to close, triggering the alarm the same way the panic button does if it is the normally open type. If the panic button is of the normally closed type, wire the "COM." and "N.O" relay contacts in series with it instead.

More exotic sensors include sensing screens to protect jalousie windows, seismic sensors that are buried and respond to footsteps or vehicular traffic, and vibration pickups for fences. The underground

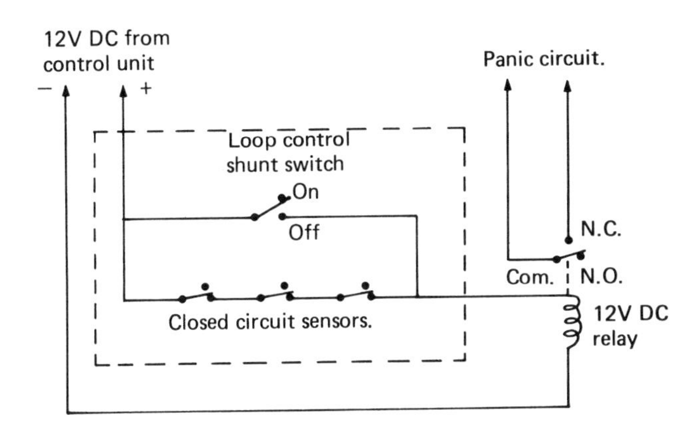

Figure 5-3. Inexpensive interior backup circuit. Only parts within dashed line are external to alarm control box.

and fence vibration sensors are useful for forming an early warning perimeter around large, fairly isolated properties. They are expensive however, and as they cannot distinguish friend from foe, you will not want them triggering your alarm system every time a kid jumps your fence to retrieve a baseball or the mailman comes up the driveway. Such sensors are best used with a local annunciator to alert you to the presence of someone on your property when you are at home. When away, you may have them turn on floodlights and a bell that operates briefly to make the person aware that their presence has been detected. If they were approaching with burglary in mind, you can be sure they will think twice. If innocent, no harm is done and no false alarm results. Information on these sensing devices is available from alarm equipment distributors and retailers. Microwave and pulsed infrared systems may also be used for outdoor perimeter sensing.

Whatever sensors you choose, be sure to follow the manufacturer's instructions on installation. Always ask yourself how you would try to disable the device if you were a burglar, and remedy any vulnerability found before someone else finds it.

WIRING

For maximum reliability and safety from tampering all wiring should be concealed, preferably within walls, baseboards, and the like as much as possible. Local electrical codes must be observed with 115V wiring, but more freedom generally exists regarding the wiring for sensors and alarm devices that operate at low voltages. Ordinary two conductor hookup wire or "lamp cord" is usually adequate and can be bought in spools of 100 feet or more, with one wire marked so polarity can be observed if necessary. For connections between sensors no. 22 gauge twisted-pair hookup wire is fine, but the thicker lamp cord (typically no. 18 gauge) is best for the higher currents of bells, sirens, and lights. If current in any one line exceeds 2 amps, use heavier wire such as no. 14 or 12 for runs of over 100 feet. If wire cannot be run inside the structure, fasten it to baseboards or walls with insulated staples and run it where furniture provides concealment. Low voltage and sensor wires may be run under carpets near the walls and under furniture but should never be run where there is heavy walking traffic over it.

When running wires make sure to tag each for identification and polarity where necessary to assure proper connection to the control panel terminals. An ohmmeter is a good way to check wire continuity or to find any break in a closed circuit sensing loop (see Figure 5-4).

Many alarm control units have inadequate internal power supplies that cannot deliver the several amperes needed for an effective array of alarm signalling devices. Likewise, the "dry" (isolated from internal power) relay contacts provided are often inadequate for the voltage and current of high-power floodlights or motor-driven devices. In such cases, you can use an external power source such as an auto battery kept fully charged by a "trickle charger" and control the load with a suitably rated relay that is operated by the DC output from the alarm control when it is triggered. Devices that require 115V AC may be operated by a relay in this manner, but will not work if the power has been cut off. In wiring to relay contacts, ignore the opposites rule used for sensor switches. When the relay is not energized (no voltage is on the coil) the contact between "COM." and "N.C." is closed, and that between "COM." and "N.O." is open. When the relay is energized (usually by the triggering of the alarm) the contact between "COM." and "N.C." is broken, and the one between "COM." and "N.O." becomes closed. Always make sure that the voltage is correct for the devices used and that the wiring is adequate for the current and protected from tampering. Rechargeable battery application is discussed further in the next chapter.

The alarm central control unit should be installed in a hidden location such as a closet, where you cannot be observed through a door or window operating the system on-off control. When with guests, have them wait outside while you operate it, and explain that the alarm would sound if you did not do it this way.

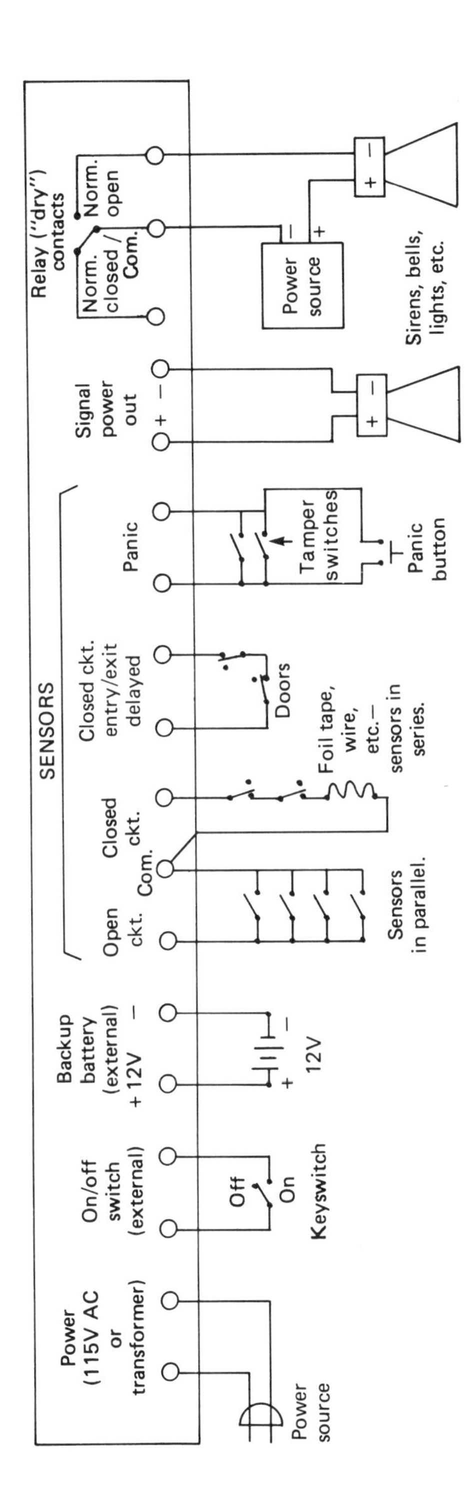

Figure 5-4. Typical alarm control unit terminal layout and connections.

It is essential that wiring be run and sensors be provided to cover all possible points of entry, no matter how unlikely. Sensors are the cheapest elements of a system, and you can never have too many.

One useful source of information on do-it-yourself alarm installation is *The Compleat Watchdog's Guide to Installing Your Own Burglar Alarm* by David Petraglia. Manufacturers' catalogs are also good sources of information. Misleading and erroneous information may be found at times, so if you find contradictions or have questions, either use your common sense or consult an expert.

BLITZKRIEG!

Let us assume that you have invested a total of $900 in a top of the line alarm control unit plus an elaborate network of sensors. When you came to the sound devices, you had only $100 left, so you settled on two big bells and mounted them inside by your front and rear doors. You had an electrician run the wiring through conduit to protect it, and the bells even have tamper switches. You feel secure.

Along comes Shiv, the 17-year-old junkie that would happily force his sister into prostitution if it guaranteed a steady supply of heroin. No brilliant dude, but he has a 9mm. automatic and a few crude burglar's tools. Given the right opportunities, $200 a day for his habit is easily within reach. Shiv doesn't go to school, but he is street smart and ambitious when necessary. Just yesterday after coppin' dope at the arcade, two members of his "crew" bragged to him of how they beat a store alarm with a can of shaving cream. Too bad Shiv never goes to school. He learns fast.

Shiv knows the alarm you have uses bells, and where they are located. It seems he knows a "friend" of your son that will tell all for a nickel bag or two, and has been in your home often enough to know it well. Shiv knows that after the mailman leaves around 1PM your home will be empty and unvisited for two hours, and that most neighbors are away too. Since your home is heavily insulated and all windows will be closed, those loud as hell's bells will barely be audible outside. He crudely forces open a window concealed from the street and neighbor's house by bushes. The alarm sounds like a machine gun attack on a boiler factory, but Shiv stays cool and once inside shuts the window. Within one minute both bells have been heavily muffled with foam from a can of shaving cream. He looks out a few windows to make sure no one's attention has been attracted, then proceeds with his business. He'll get good and high tonight.

Had one of those bells been in a suitable protective box mounted outside the house in a hard to reach and very visible place, Shiv would have gone elsewhere. Even then it is bad planning to spend $900 for a control box and sensors but only $100 for the noise, with no lights or other help summoning devices. It's like putting $50 speakers on a $1000 stereo unit. Ineffective.

Up to this point mainly sensing devices and control units have been discussed. Without effective signalling devices, all the money and effort invested in them is a waste. You might as well have spent it on insurance instead, because you'll need it.

Audible alarm devices should be as loud as possible over as wide an area as possible, both inside and outside the premises. Inside, the sound level should actually be painful, 120 db SPL or better. The number and type of devices necessary to produce this ghastly din depends greatly on the size and acoustical properties of the rooms they are in, "dead" rooms requiring more, reverberant rooms less. A good mounting spot is a corner where two walls meet the ceiling and no sound absorbent objects such as drapes are nearby. This focuses the sound like a horn and can make a device more efficient, especially one such as a bell that radiates sound in all directions. Electronic sirens and buzzers with acoustic horns are more directional and less critical in their placement when used indoors. Just aim them at doors, windows, and other likely points of entry. When used outdoors, direct them at streets where there is heavy pedestrian traffic and at neighbors' homes. If you can get some surplus factory or diesel horns, an air raid siren, or anything else capable of turning eardrums to mush, do so. An air raid siren may be overkill for a small apartment but if you have a home in an isolated area it is ideal. It is made for outdoor use and can be heard for miles. Just make sure the power connection is tamperproof, and mount it high off the ground on a solid structure.

Keep in mind that if a device is rated as producing a 95 db sound level, two such devices together produce 98 db, not 2 × 95 db. Each 3 db increase requires a doubling of power. A 110 db noisemaker would require eight 101 db devices to equal its sound output. Go with a few of the loudest units you can get rather than several less loud ones that will wind up costing much more for the same level of effectiveness. Devices rated at less than 110 db at ten feet are best avoided, as are ones that give a high db figure but don't mention the standard ten foot distance. Such specs are misleading.

Bells are probably the noisemakers most vulnerable to tampering. If you use them, be sure they are in good protective enclosures and mounted well out of reach. Shiv is watching you.

People tend to ignore a constant pitch siren because there are so many false alarms from auto systems. A warbling electronic siren or a horn that blasts on and off repeatedly is more effective. Electronic noisemakers that use horn type loudspeakers are more efficient than most motor or bell mechanism driven units and therefore more of them can be powered by a typical alarm control. Electronic sirens are preferable to bells, but even though they are more tamper-resistant, they must be placed well out of reach. Tamper-proof enclosures are desirable and are a must for outdoor use. Self-contained electronic sirens have the electronic driver (signal generator) and horn speaker joined together in one package. Also available are separate speakers, several of which may be driven by one high power driver. This is an efficient and effective arrangement for a large home or apartment.

You can spend a thousand dollars on the best electronic sirens and bells, put them into locked steel enclosures complete with tamper switches, mount them so that they can only be reached with a crane, and still have them quickly rendered useless by a smart burglar with a wire cutter. All wiring from the control unit to the signalling devices MUST be protected every inch of the way. This means running it in electrical conduit or using flexible metal-shielded cable so that it is protected and looks like ordinary house power wiring.

Added security for the signalling devices may be had by running two or more independent circuits for them from the central control. Let's assume you are going to use four electronic sirens, two inside and two outside the house, plus a rotating police light in an upper story window. Using the information supplied with these devices or available from the manufacturers, add up the total current required. In this case, we find the sirens take 1/2 ampere each and the light 2 amperes. Our goal is not to evenly balance the current load between the two separate circuits we will use, but to wire the devices to them so that if one of the two lines fails, the other will provide the most effective possible protection. If the four sirens were on one line and the light on the other, the load for each would be equal, but should the siren line be defeated, the light, which operates silently, would be of little value. In this case we are better off with the light and two of the sirens, one inside and one outside the house, on one circuit, and the other two sirens, also one inside and one outside, on the other. Now the failure or defeat of any one line leaves at least two sirens, interior and exterior, still functioning. We now have two pair of load wires running from the control, one with a 3 ampere load (two sirens plus the light), the other with a 1 ampere load (the other two sirens). Now put *fuses* in the lines; a 4 amp slow-blow or 5 amp regular fuse for the 3 amp load line, and a 2 amp slow-blow or 3 amp regular fuse for the 1 amp line. In the event of a short circuit on one line—accidental or deliberate—the fuse will protect the alarm power supply from damage (assuming it is properly designed and has sufficient capacity), quickly disconnect the shorted line and permit the other to continue functioning. In all these circuits, the load devices are connected in parallel and the fuse is in series with one of the two wires that go to them. The fuses and fuseholders, available from electronic parts suppliers, should be contained in the locked control box or otherwise made inaccessible to possible tampering.

One nice trick is to get a cheap bell or siren speaker to use as a decoy. This is deliberately placed where it is easier to tamper with than the real sound devices; it is not connected to the alarm output but is wired instead into the closed circuit sensing loop. Use exposed wire to connect it, but don't make the trap too obvious. When the clever thief cuts the wire, he'll learn the hard way that he was not clever enough.

Another good decoy is to put alarm keyswitches in or near your doors, even though your system has entry/exit delays and is operated solely by a control inside your home. Use cheap, easy to defeat keyswitches for these, wired into the alarm sensing loops so that the burglar's operating them sets off the alarm. Imagine the look on his or her face when after several minutes of nervous lockpicking, success turns to surprise.

Lights are best used in creative ways as a deterrent, making your home appear occupied when it is not and illuminating dark areas that can conceal prowlers. An alarm with lights should focus intense illumination on the likely points of entry, plus create an unusual visual signal such as a flashing strobe or rotating police light in a window or on the roof. To police in the street the noise of an apartment alarm may be hard to pinpoint. If your window is lit up like a disco they'll know where to go, plus where to look for the body of the burglar that fell from the fire escape when he was scared witless.

It has often happened that although an alarm is blaring loudly, no one calls the police. For effective security a means of quickly summoning assistance is a must. Most reliable, but most costly, is the automatic notification of a central station alarm service. A less reliable alternative is a telephone dialing device that is operated by the alarm and sends a recorded message to the police and/or others. A network of neighbors can be organized to respond to the sound of an alarm or an intercom or radio transmission it sends when triggered;

the reliability of this depends on the number of people involved and the degree of their concern. Where there have been several crimes on a particular block or in a given building, there are usually enough such seriously concerned people to make this system fairly reliable.

The simplest arrangement is to have an agreement among at least four or five households that ALL will call the police IMMEDIATELY upon hearing an alarm. They then should go to the street, if necessary, to guide the police to the proper house or apartment, but should never investigate themselves. If five homes participate in this, the likelihood that at least one is occupied at any given time is very good, and if many are home, several calls to the police are better than one.

Another arrangement that Mr. Petraglia suggests in his book is a bit slower in calling the police. First have neighbors call the home or apartment from which the alarm is heard. If there is no answer, a persistent busy signal, or a stranger answers, they should call the police. If the resident answers and says it was just a false alarm *but fails to give an unusual prearranged password* they also should call the police, as this will indicate that something *is* wrong, such as they are being robbed at gunpoint. In such a case be sure to tell the police of this possibility.

Needless to say, false alarms will quickly destroy the spirit of neighborly cooperation; all the more reason to avoid cheap motion detectors, sound sensors, and similar electronic devices that are likely to cause false alarms.

Automatic phone dialer units triggered by an alarm are not too reliable even when they do not malfunction. Because of frequent false alarms (usually not the dialer's fault), police give such calls low priority, and in many urban areas dialers are forbidden altogether on the police emergency line. It is possible to defeat many such devices by cutting phone lines, physically damaging them, cutting off their power, and other means. If you have such a unit, consult your local police on how best to use it; it may be better to have it call someone other than the police. It should not be used to call an ordinary answering service though, as such services are not equipped for security work and their personnel are not trained for it. Phone dialers are not particularly recommended, but if you want one, here are the features to look for:

* A backup battery and the ability to continue functioning if either the plug is pulled or the alarm system triggering it is cut off.

* Redial and multicall capability (redials if number is busy and can make several different calls in succession).
* Line seizure and release to defeat tampering attempts.
* High voltage circuitry protection.
* Internal speaker for testing.
* Enclosed in a strong key-locked cabinet that can be mounted fully concealed.
* Key-operable abort switch to be used if the alarm is accidentally triggered.

Be sure that you disconnect the dialer from your phone line when doing routine testing and reconnect it when done. A good taped message format is this: "This is an emergency recorded message. A burglary is in progress (or a fire has been detected) at (full address, including apartment number if a multiple dwelling). Please send the police (or fire department) immediately. The address again is (repeat)."

As the telephone may be your only link with the outside world in an emergency, integrity of the line should be considered an important part of overall home or apartment security. On some phone dialers a line monitoring feature sets off your alarm if the phone line is cut. The dialer will not be able to send any messages of course, but the alarm may scare off the intruder and will alert you if you are at home. Such a feature is *NOT* desirable, however, if your local phone lines are not highly reliable; line failure from causes other than the deliberate cutting of the line can cause a false alarm. A bug unit is available that connects to your line and allows you to hear any sounds in its vicinity when you call from an outside phone. Of course this must be disconnected when you are home to ensure your privacy. Some high-security vaults are equipped with such devices so that they can be monitored from a distance.

If you have an emergency phone dialer or a phone line used by your alarm system, the fact is best kept a secret. If the phone company knows of it, they may mark your wires with tags or sleeves if you live in a multi-dwelling building. This is for their convenience, but it endangers your security. Find the junction box or wireroom in your basement, garage, or meter room. Wires are usually marked by apartment or room number. If yours have any tags or sleeves distinguishing them from the others, remove them. They are more useful to smart thieves than to the phone company, and you have a right to your privacy and safety. You might also wish to tag your phone line "LEASED LINE." This indicates to the criminal that guards or the police may be summoned if the line is cut, a good reason for him to go elsewhere.

CENTRAL STATION SERVICES

The central station is a "nerve center" at which many communications channels from individual alarm systems are monitored twenty-four hours a day by specially trained personnel. When an alarm signal is received, appropriate action is taken immediately; this can range from merely calling the police or fire department to the dispatch of armed guards to the premises. To be safe against defeat, the company must have a highly secure base of operations, state-of-the-art equipment, and well-trained, carefully screened personnel. Unfortunately, many such services are lacking in one or more of these, and as a result are worthless or worse.

The least expensive, and by far the least valuable central station services do little more than sell you an alarm with a phone dialing device. When the alarm is triggered it calls their office, and they then call the police for you. This is no more reliable than their hardware, their personnel, and your ordinary phone line, which in most cases can be easily disabled. Unlike quality services, they do not use a monitored high-security ("leased") phone line to your premises, and the police response may be more rapid if you can call them directly with your own equipment. Where police-calling dialers are prohibited however, a good service of this type has some value. As with most other things you will get what you pay for, and less if you are not careful; do not expect a real guard service for a mere $25 per month. You can expect to pay over eight times this figure for a leased line alone!

If you want a service that responds to your burglar alarm with armed guards, you must choose it very carefully and be willing and able to pay for it. If you live in a isolated area or have valuables likely to attract professional burglars, you should have such a service. If you cannot afford it though, you probably don't need such a high level of security. Professional burglars usually hit only the wealthy.

Essential to such a security service is the integrity of the communications channel to it, whether telephone or radio. A leased (used for no other purpose) telephone line between your premises and the central station is constantly monitored by a sophisticated electronic system to detect if the equipment is faulty or has been tampered with. In the event of an alarm, the central station receives a coded message that tells if it is due to burglary, fire, holdup, or any other condition it is designed to indicate. The time and nature of the alarm are recorded, "runners" (guards) are sent out, and the police are called in the event of a burglary or indication of tampering.

The information in Chapter 4 regarding private guard services also applies to services that respond to a central station call. The well-known companies that have been in business for many years are generally reliable, but some others are not. As with alarm installers and other guard services, you should check the firm's reputation with the police, Better Business Bureau, consumer protection agencies, and the National Burglar and Fire Alarm Association. Also check with neighbors who use the service, especially any who have had experience with the service's response to an alarm. The guards should arrive within fifteen minutes of the alarm and never take much longer than that. The firm should provide you with a written explanation of the services guaranteed for your fee, and carry liability insurance for your protection from lawsuits because of action they may take on your property or in response to your alarm. Their guards should be required to pass a background check prior to employment, be well-trained, and bonded.

Many central station services provide and install the alarm hardware, which you can buy or rent. Rental is best, unless you want to do your own maintenance or expand the system. Such a service should be qualifed to do UL certified work, and be able to issue a certificate of compliance. They may request that you give them keys to use when responding to an alarm; you should only do this if your contract specifies that they may not enter your premises without the police present. Make sure that they have your business or vacation address and phone number, and will inform you immediately of any emergency.

The guard service should not be given a key to deactivate your alarm, only you should be able to do this. If you will be away for several days, you may leave an *unmarked* alarm key with a trusted relative or neighbor who will keep it in a secure place and allow guards to use it without taking possession of it. This should only be a level-headed person you would trust with your life, and they should be instructed not to let the guards know they have such a key until they are certain of their identity. Crooks can put on uniforms too. If suspicious, this person should call the police.

Underwriter's Labs has a grading system for central station services and alarms, ranging from AA to C. AA is most secure, C is a minimum requirement. While a lack of UL certification is not proof of inadequacy, incompetence, or dishonesty, you are far less likely to go wrong by choosing a company with such credentials. Again, use extreme caution in selecting such a service. There are some around that would hire Shiv, no questions asked.

WORDS TO THE WISE

Packaged systems are offered by many alarm dealers. Some are satisfactory, but many lack desirable features and adequate signalling devices. These systems should be carefully evaluated and not purchased if they are incapable of being improved upon by the addition of more sensors and sound, light, and help-summoning devices.

"We have met the enemy, and he is us." Without every member of your household thoroughly trained and drilled in the proper use of your home security system, this is literally true. If one individual is careless and leaves home without using the alarm "because I'm coming right back," that is all it takes for an observant criminal to gain entrance undetected. If you do come right back, you may lose your life as well as your property. It is important that children be taught never to reveal home security details to anyone, and teenagers should never leave acquaintances alone in the house. Your alarm system should be frequently and thoroughly tested. If it suddenly becomes defective, consider the possibility of sabotage and do not leave the premises unguarded until it is repaired. If there is evidence that the system has been tampered with, immediately change and improve the system to make it invulnerable to future attempts.

I have avoided recommending specific equipment brands and manufacturers, as every year new equipment is offered and older models are removed from the market. Manufacturers and service companies come and go, and the profit motive makes an unbiased opinion hard to find. Be wary of fads and heavy or high-pressure advertising, and give an immediate and unconditional *NO* to anyone who uses scare tactics. The recommendations of publications that carry paid advertising or are put out by someone in the business should be read with skepticism. See the Yellow Pages under "Burglar Alarm Systems" and industrial directories such as Thomas' Register for manufacturers, distributors, retailers, and installers. Many free catalogs are available that have technical information and prices for specific products. Being listed in a directory is, of course, no guarantee of competence or integrity, and otherwise impressive catalogs may contain some inferior products. Check objective evaluations such as those of *Consumer Reports* and *Consumers' Research*; their findings are far more reliable than any advertisements. These and other sources of information are available free at your public library. A security system is not just a major purchase, it is an investment in the protection of all that you own, including your life. Study the market well and choose wisely. No one on this earth can protect you, your loved ones, and your property as well as you, as no one else cares as much as you do. Just use your intelligence and common sense, and beware of anyone that tells you or infers that you lack them.

6

BUILD THIS RELIABLE LOW-COST ALARM

The protection afforded by an electronic alarm system can be had at a much lower price if you are willing and able to build and install it yourself. If you or someone you know and trust have some skill in electronics, the construction and installation of the unit given here will provide a very reliable no-frills system at minimum cost (see Figure 6-1). Features are:

* Hard wired perimeter system.
* Provision for both open circuit and closed circuit sensors.
* Backup battery kept fully charged by 115V line.
* 12 V DC for signalling devices with a current capacity of 5 amps.
* Panic switch.
* Dry relay contacts available.
* The added security of doing your own private installation.

HOW IT WORKS

Switch S1 is a high security key-operated switch mounted in the front door or door frame; S2 is an indoor control mounted in a well-concealed location. Either of S1 or S2 when switched on connects the 12 VDC from the battery and its charging supply to the alarm circuitry. A small current (about 0.2 ma) flows through R3, R4, and the closed circuit sensor loop. The voltage on the gate (terminal 3) of

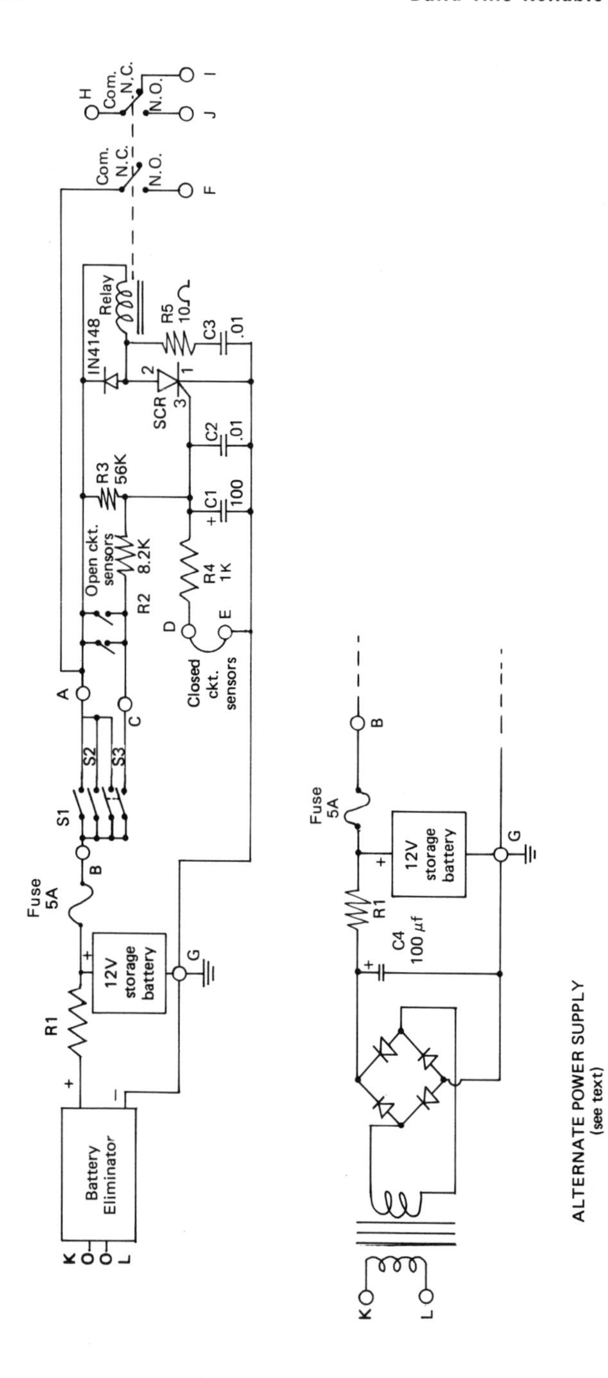

Figure 6-1. Schematic diagram.

the SCR is about 0.2V, which is not enough to trigger it. Should a closed circuit sensor open, the voltage rises and the full 0.2 ma goes into the gate, triggering the SCR. Once triggered the SCR conducts current that causes the alarm relay to operate, and it remains in the conducting state until the alarm is turned off with the switch. If instead of the closed circuit loop opening, an open circuit sensor contact is closed, a higher current (about 1.4 ma) through R2 causes the voltage at the gate to rise enough to trigger the SCR. The double pole panic switch connects the battery to the system and triggers it at the same time; the alarm sounds immediately and continues until all switches are opened.

Capacitors C1, C2, C3, and resistor R5 act as filters to prevent false alarms due to radio or other signals picked up on the system wiring. The diode (1N4148) across the relay coil prevents a high voltage "spike" that might damage the SCR when the alarm is turned off. The 5A fuse protects the system from damage in the event of a short circuit. Without this fuse, the current resulting from a short could be high enough to cause a fire. The 5A fuse is an essential part of the system and should *NEVER* be bypassed or replaced with a fuse of a higher current rating.

PARTS LIST AND INFORMATION

1 Relay, 12 VDC coil, 2PDT contacts rated 5A min at 115V AC.

1 SCR, General Electric C106F 1 or 2, C106A 1 or 2, C106B 1 or 2, C106C 1 or 2, or C106D 1 or 2.

1 Diode, 1N4148 or equivalent.

1 Fuse, 5A slo-blow rated at 25 V min with fuseholder for mounting on circuit board or within the enclosure.

1 2PST toggle switch (panic switch).

2 Switches suitable for ON-OFF controls, one or both key-operated.

1 Capacitor (C1) 100 uf 5VDC min.

2 Capacitors (C2, C3) .01 uf disc 25 VDC min.

1 Resistor (R2) 8.2K 1/4 w 5%.

1 Resistor (R3) 56K 1/4 w 5%.

1 Resistor (R4) 1K 1/4 w 5%.

1 Resistor (R5) 10 ohms 1/4 w 5%.

1 Terminal strip, 12 or more terminals, insulation rated for 115V AC minimum. (See Construction and Installation below.) Alarm devices rated for 12 VDC with a total current requirement of 5 amps or less.

Sensing devices.

12V rechargeable battery and power supply. (See below.)

Suitable hardware, wire, tools, and construction materials.

The prices charged the public by electronics parts retailers vary from high to outrageous, with a mere 1/4 W 5% resistor sometimes going for as much as thirty cents. In spite of this, you still save considerably by building this alarm yourself. If you or someone you know works in the electronics field and can get parts and materials from a distributor or through a company purchasing department, you can save as much as 75 percent. If not, it pays to seek out dealers in surplus parts and equipment, rather than go to the nearest store.

The selection of switches and the sensing and sounding devices should be guided by the information given in Chapter 5. By constructing and installing the system yourself you save a great deal of money, and some of your savings should be invested in more and better sensors and sound devices to ensure greater invulnerability and effectiveness. While this control unit lacks some of the fancy features found on more expensive commercial products, it can provide reliability and performance comparable to or exceeding theirs.

An automobile or motorcycle battery is ideal for this system—even a used one that has trouble starting a car in very cold weather is satisfactory as long as it is not damaged and does not have a "dead" or internally shorted cell. Just clean it up, add water if needed, charge it fully, and have it tested under load if any doubt exists as to its condition. Even a marginally good battery has more than enough capacity for this alarm. Physically smaller ni-cad, lead-acid, or "gel-cell" rechargeable batteries may be used, but most will quickly run down under a 5 amp load, if they can deliver that much current at all, and ones with high capacity are expensive. The car battery is the cost effective way to go, and the motorcycle battery is a good physically smaller alternative. Smaller batteries should be used only if the load will be less than 1 amp or so. High-power alarm loads, such as 115V floodlights, should be operated through the dry relay contacts as shown, but keep in mind that these will not work if the 115V is cut off. Only the devices powered by the battery are power-failure-proof. Be sure not to exceed the relay contact ratings; if desired, a relay with more contacts or higher ratings may be used.

There are several ways of making the power supply that trickle charges the battery. This supply is permanently connected to the 115V line and provides a constant current when the voltage is present. Should the 115V fail or be cut off the charging stops, but the battery, normally fully charged, maintains the small current required by the alarm. Unless the alarm is triggered, the battery can power it for months; if it does sound, a car battery will deliver an amp or two for days or at least many hours. Should the battery ever be completely drained, it must be charged with a high-current charger before being put back into the alarm system, as the trickle current would take a very long time to adequately charge it.

One power supply possibility is to use a "battery eliminator" with a rated output of 12 or 18 VDC. The 12 VDC units are the most common, but some may not produce a sufficiently high voltage to force a 5 ma trickle charging current into a fully charged 12V storage battery as is required. Units rated for 12VDC at 100 ma or more will probably work. To be sure, put a milliammeter in series with a resistor substitution or decade box between the positive output of the eliminator and the (+) terminal of the fully charged battery, connect the negative output to the (−) battery terminal, and adjust the resistor for a current of 5 ma. Start with a value of about 2.2K; if the resistance must be brought below 270 ohms to get the 5 ma, the output voltage of the eliminator is inadequate. The figure of 5 ma is for an auto battery, if a smaller battery is used, the current must be lower. A good rule is 1 ma for every 2 ampere-hours of battery rating up to a maximum of 5 ma, but it is advisable to check the manufacturer's specs for constant current charging for any given battery. For maximum battery life stay well within their ratings, as excessive constant current can ruin a battery quickly. If the eliminator can be used, select the value of R1 to obtain the proper current.

If a suitable eliminator is not available, use a step-down transformer with a 115V 60HZ primary and 12.6V secondary. You can use a full wave bridge rectifier made with four diodes (1N4001's are suitable) and a 100 uf 25 VDC min. capacitor (C4) as shown in Figure 6-1. Here also, R1 is selected for the proper current into the battery when it is fully charged. If the secondary voltage is higher than 12.6, use 1N4002 or 1N4003 rectifier diodes and a 100uf 50 VDC min. capacitor. If the voltage is above 35V, the transformer may not be suitable, although if the secondary is center tapped, you can use a two-diode rectifier arrangement with the center tap grounded. As the supply delivers a maximum of 5 ma, which amounts to only roughly 100 milliwatts, the transformer can be one of low power rating.

As this alarm requires a keyswitch in an outer door, it is essential that it be one with a high-security cylinder protected by a guard plate. Also advisable is a tamper switch that will sound the alarm if the cylinder is physically forced. This can be devised by using a length of foil tape or very fine wire electrically insulated from, but physically connected to, the cylinder. Wire this in series with the closed circuit sensor loop so that if the cylinder is twisted or pulled the conductive path is broken and the alarm sounds.

In spite of protective hardware and a tamper switch, the keyswitch outside the premises is still vulnerable to the highly skilled burglar. Good additional protection may be had by adding a backup switch or other contacting device in a cleverly concealed location. In this case, this switch must be connected in *parallel* with the main keyswitch. The wires from the hidden device should not go directly to the terminals of the keyswitch, but should connect to its wires at some distance from it. Either of these switches can be used to turn the system on and off, but *both* should be used at the same time. As long as either one is on, the system will be on, so the burglar who can defeat the main keyswitch will still not be able to turn the system off unless he knows of the hidden switch as well. Use your imagination in devising the backup switch arrangement; you want to be able to operate it inconspicuously. To give plans for such an installation here would only tell burglars what to look for and how to beat it. Secrecy is often essential to security.

The circuitry as given here lacks an automatic cutoff/rearming feature. If you wish to add this, use a time-delay relay that is supplied with 12 VDC when the alarm sounds, and after five minutes or so opens an isolated, normally closed relay contact. Wire this contact in series with the DC line to the alarm relay. In Figure 6-1, this contact goes in the line between the top of R3 and the 1N4148 diode. This will momentarily break the circuit, allowing the SCR to reset. If the alarm was triggered by a sensor on a door or window that was restored to its normal condition by the closing of the door or window after the alarm sounded, the SCR will remain reset, the alarm will cease to sound and will be rearmed. If, however, the sensing device remains in the triggering condition, the alarm will continue to sound until it is restored to normal. Without an automatic cutoff/rearming system, the alarm will continue to sound until either it is turned off or the battery runs down. A drained battery and annoyed neighbors are a small price to pay for the prevention of the loss of your property to a criminal. If you are going to be away from home for an extended period of time and you have a very trustworthy friend living nearby, you may wish to leave an unmarked key for your alarm with him or her. Should your alarm sound, he should call the police immediately; most likely they will find that the intruder left quickly when it sounded. If a door or window is open, your friend can close it and reset the alarm by turning it off and then on again. He should make sure that nobody is watching him do this, lest they try to steal the key. He should also be certain that everything is left locked and the alarm is on. Should there be any problems he cannot handle, he should have a phone number where you can be reached. A shortened vacation is better than coming back to an empty home.

CONSTRUCTION AND INSTALLATION

A sturdy enclosure, preferably a locked steel box, should be used to house the battery, power supply, and other circuitry. Ventilation holes are important to allow the escape of any corrosive gases the battery may produce, so that the enclosure and circuitry are not damaged by them. Be sure also that the battery is securely mounted so that its positive terminal cannot come into contact with the metal box. Only the ground point of the circuit (G) should connect to this enclosure. All circuitry shown in Figure 6-1, except for the sensors and outside control switches, should be mounted in the box, and the fuse must not be accessible without opening it or else the alarm could be disabled just by removing the fuse. The small components can be assembled on "perf-board" and should be securely soldered; a PC board is unnecessary. Heat sinking the SCR (see Figure 6-3) is not necessary, but be sure that the metal tab does not contact any other circuit point, as it is connected internally to the anode (terminal 2). The terminal strip mounts on the outside of the enclosure and should be the type whose connecting pins project back through holes in the enclosure for wiring to the circuitry. After all wiring is complete and the system has been tested, the terminals should be covered over to prevent the danger of shock or a short circuit.

Use Chapter 5 as a guide for installation and wiring. Figure 6-2 shows how to hook up all external components, including the use of the relay contacts for 115V alarm devices. If these are not used, make no connections to terminals H, I, and J. Alternatively, these contacts may be used to operate an alarm accessory such as a transmitter or phone dialer.

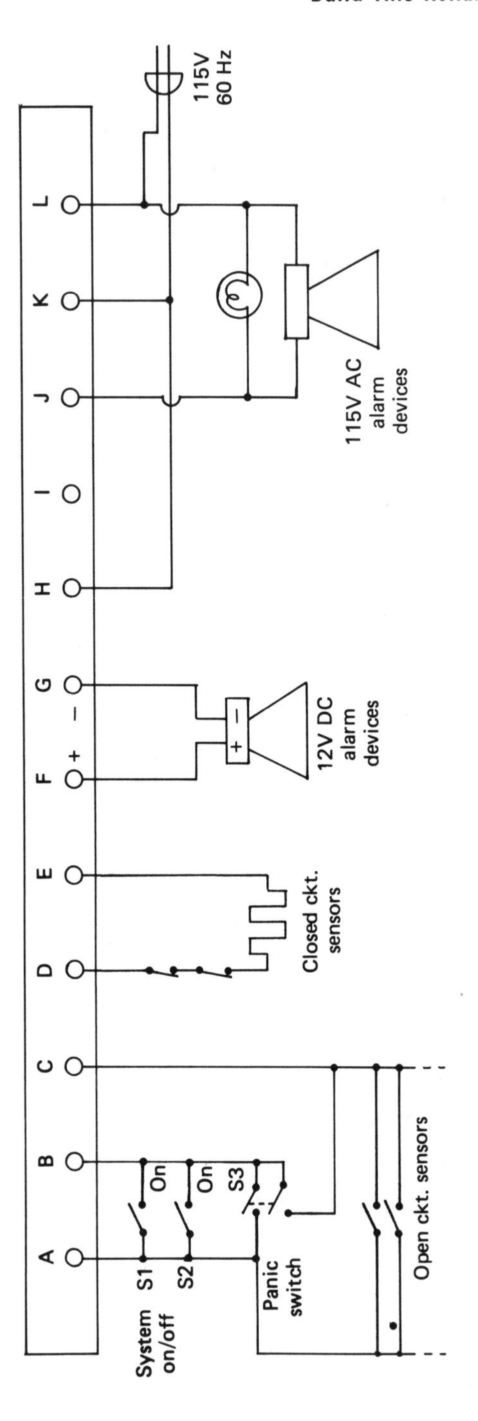

Figure 6-2. Terminal strip.

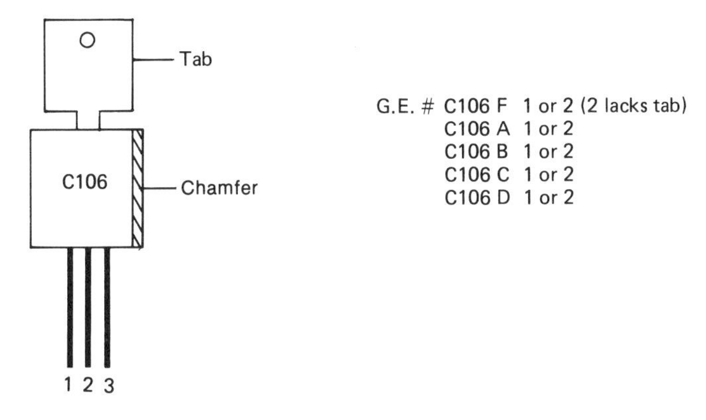

Figure 6-3. SCR.

S1, S2, and one pole of S3 carry the full current of all the alarm devices, which can be as much as 5 amps. In wiring them, no. 18 (lamp cord) wire may be used if the length between switch and control unit is less than ten feet; if it is over ten feet but less than fifty feet, use no. 12 wire. Runs of over fifty feet are not recommended unless the maximum current is lower. This also holds true for the wiring between the control box and the alarm sounding devices.

This alarm is not only an enjoyable project, but one of great value. Perhaps when the human race gets its head together such devices will become as obsolete as a medieval suit of armor. Until that long awaited day, use it well.

7

STASH IT AWAY

Merely hiding, camouflaging, or locking away possessions to prevent their theft is almost as defeatist as planning what to do if you are robbed, but not bothering to lock your door. Such concealment is a sensible precaution in the case of small or medium-sized objects of high value, but the greatest effort and expense is best put into making your house or apartment as burglarproof as possible. Once you have done that and done it well, you can consider the remote possibility of a successful burglary and take steps to minimize your losses.

By concealment we in effect prevent the burglar from knowing of the existence of a desirable object. This is far better than saying "Here it is, but you can't have it." "Wanna bet?" thinks he, and if he wants it badly enough, he will find a way or die trying. A low profile is the best concealment; if you let your ego control you and take a superior, snobbish, or exhibitionistic attitude by virtue of wealth, social status, or just plain pretense, you can be sure you will attract thieves and make enemies very quickly.

CONCEALMENT: CONVENTIONAL AND UNCONVENTIONAL

What have you got to hide? Basically, most burglars seek cash, checks, credit cards and ID's, securities, jewelry, weapons, furs, tools, cameras, watches, TVs, video equipment, audio equipment, calculators, silverware, typewriters, collectibles, musical instruments, drugs, computers, and bicycles. Illegal possessions—weapons, drugs, and so forth—are among the thief's favorite items; they command a high price on the black market and the loss will not be reported to the police. Criminals seek value, portability, and marketability and shy away from one-of-a-kind or customized items or anything that may be identifiable as stolen property. Many of their favorite goodies are found in very predictable places, and for this reason the ransacking thief hits the bedrooms first, and then the living room. Unless he or she has time to spare, he will generally ignore or at least not be very thorough about the bathrooms, hallways, or the kitchen, although he is likely to check the most common hiding places in these. Time is his worst enemy and the sooner he escapes the better, unless he is unusually confident or brazen. Only the professionals will have the crime so thoroughly planned that they can be certain they will have a good period of time with little possibility of being discovered in the act. Knowing this typical modus operandi is valuable in planning your defense. Keep in mind that it is NOT true that a burglar will not return for a second helping; if he did well once, he is likely to come back to the same premises again and again.

To present a list here of the many possible hiding places or methods of concealment would be counterproductive, as the better known such information is, the lesser its value. Burglars know and usually check the common hiding places. Hiding places to AVOID are these: kitchen jars, cabinets, cans and bowls, laundry hamper, medicine chest, refrigerator and freezer, in or behind drawers, picture frames, mattresses, in or under chairs or beds, boots or shoes, air ducts, curtain hems, and garment pockets.

Thieves watch the market for the latest concealment devices and ideas and know where to look for them, so steer clear of commercially available gadgets such as phony electrical outlets and use your own ideas. For convenience, we often keep some of the burglar's preferred items such as cash, checks, credit cards, weapons, and jewelry, in just the places he will look for them. Leave a few dollars in these ordinary places to pacify him, but hide the bulk of your valuables in the most unlikely places imaginable. The best places are those that require the most time and work to find and open, especially ones that require tools to open. Some old ideas are useful under certain circumstances; hiding things in a hollowed out book is risky if you have a dozen books on a shelf but safer if you have several hundred. Make sure however that the book will remain closed if it is thrown to the floor. Concealment behind a baseboard molding or under a rug is safer in a large home or apartment than in a small one. Unless you do a bad job of it so that the hiding place is visible, no burglar is going to spend hours pulling up all the rugs and baseboards. Some of the best hiding places are well-constructed hidden compartments in walls, ceilings, and floors. Some people have hidden small valuables concealed among foodstuffs but that can be hazardous.

If you have a variety of cleverly chosen hiding places, the possibility exists of forgetting where you put what or even inadvertently discarding something of value. An up-to-date home inventory list should be prepared, but it must be kept in the most secure hiding place of all, so that it doesn't wind up being a burglar's treasure map. Off the premises concealment such as a safe deposit box is best for such records. If you have a domestic employee, it is possible that some of your hiding places are known, so be sure to change them—and your list of what is where—should such a person quit or be fired. Only the closest and most trustworthy adult member of your family should know of the existence and location of any hiding places in your home. Also worth listing are the model and serial numbers of theft-prone property such as TVs, stereos, and other appliances. This will make their recovery more likely if they are among stolen items seized by the police.

A good idea is to equip your closets with good doors, pinned hinges (See Chapter 2), and key-operated locks (hide the keys well!), keeping valuables such as furs or weapons in those closets where they would be least likely to be found. Working against the clock, the intruder may not bother with the closets at all, and if he breaks into one and finds little of value, he'll probably pass up the others as not worth the time. Such doors and locks on bedrooms are good also, but they must be easy to open from inside without a key. Anything that will slow him down works in your favor.

Custom-made vault doors for a room or closet are available, as are steel roll-up closet doors that can be installed by the handy home or apartment owner. These are only of value if the strength of the locks, frame, and surrounding room or closet walls are at least equal to that of the door. A good safe, concealed within a closet so protected, is truly high security.

A good theft deterrent for larger items such as cameras and audio and video equipment is the deliberate "customizing" of such items by defacing them with paint or other difficult to remove material that makes them readily identifiable. This is not practical, though, if you wish to be able to resell such property yourself, as it decreases the market value.

Anything that can be done with large items to make their theft more time consuming or make it necessary to damage the item to get it is a good deterrent. Furniture may be bolted to the floor, and the power cords of appliances or stereo equipment may be tied together in a huge jumble. Bolting down is particularly recommended for outdoor furniture and accessories left on patios or in recreation areas.

Strongboxes are often far from strong, and even a good one hidden in a drawer or closet is a poor bet as if it cannot be opened quickly the whole box will be taken. A good ruse, though, is to keep a well-locked box filled with rocks or sand where it can easily be found;

imagine the look on the thief's face when he finds that the rocks he stole are not the kind he was hoping for.

A safe deposit box is one of the most secure places available, in spite of the fact that some have been successfully burglarized. If this worries you, insurance is available; banks themselves will take no responsibility beyond routine security precautions. The biggest problems with such boxes are limited space, limited access, escalating rental fees, and the uncertainty of access in the event of the financial failure of the bank or other disaster. This is the place for all valuables that are not required to be in the home such as stocks, bonds, deeds, coins, stamps, insurance policies, jewelry not regularly worn, and most important, your inventory list of possessions including serial numbers and identifying characteristics. Photos of property and receipts can also be kept here, and in the event of fire or theft, they may be your only proof of ownership that is acceptable to an insurance company or the IRS. If you do not trust banks, there are private safe deposit box and vault companies that can provide greater privacy and security. Check them out carefully before utilizing their services.

Burial as a means of concealment can be effective, but has several risks and drawbacks. Only valuables that are not susceptible to damage by moisture should be buried, as even good waterproof containers can leak, corrode, or decompose in time. Plastic containers such as film cans with tight lids are probably best for this purpose. Other risks of burial are the possibility of being observed while doing it, the potential of discovery with a metal detector or by accident, and the possibility of forgetting exactly where the burial was done. A carefully drawn map showing the location of the spot and a list of the concealed items should be made and kept in your safe deposit box or an equally secure location. Buried treasure that neither you nor your heirs can find is lost just as surely as if it were stolen.

SAFES

There are two types of safes, fire resistant and burglary resistant. Fire safes can be cracked by an amateur with ease and should never be used for protection from theft. Even the best safes can be cracked by a pro given enough time, and when improperly secured, can be stolen whole and opened elsewhere. A good burglar resistant safe must be properly installed and well concealed to be effective. If you let others know that there is a safe in your home, word may reach the ears of someone who enjoys a challenge. If possible, avoid leaving any records of your purchasing it. Silence is golden.

The cost of a safe is tax deductible if it is used for the protection of valuables that are to be sold later at a profit, but of course taking such a deduction reveals to others that you have a safe. The same is true of taking a deduction for a safe deposit box rental fee; privacy may be worth more than the deduction.

Wall safes are the cheapest, but by far the least secure. The best wall to put a safe into is thick reinforced concrete, and the safe should be in a well concealed location, not merely hidden by a picture or mirror. The average burglar is not a skilled safecracker and is likely to leave even a weak safe alone when he discovers it, but he will probably return with someone else in the future. An even worse possibility is that armed robbers will show up and force you to open the safe for them.

Far better than a wall safe is one that is embedded in a thick concrete floor or is at least secured thereto with bolts whose heads are inside the safe. No matter how heavy it is, a safe should never be left on wheels or skids; if you expect a safe to stay put with no bolting to the structure, it had better literally weigh a ton, preferably more. Minimum requirements for safe construction are a half-inch thick solid steel body and a one-inch thick solid steel door constructed with inaccessible hinges. It should be new (never buy an old, used, or repaired safe), have hardened steel plates over the lock mechanism areas for drill resistance, and a tamperproof relocking mechanism backing up the main lock. Stainless steel is better than ordinary steel for safe construction but is expensive. A desirable feature is built-in alarm system heat detectors and a heat-conducting copper lining.

Combination locks are considered by some people to be more secure than key locks for safes, but the combination should be changed from time to time. Memorize the combination, and record it in some form of code on a slip of paper to be kept in a secure location, preferably off the premises. It is possible for the combination to be learned by watching you open the safe even from a distance, so be sure to conceal the dials with your body when you operate them. After closing the safe, make sure to spin all the dials several times. If the safe has key locks at least two requiring different keys should be present, and one of the keys should be kept concealed off the premises.

You should never buy a safe that has not been evaluated by Underwriters' Laboratories and does not bear their rating label. The lowest burglar-resistance rating is TL-15, which means that the safe can resist attack with ordinary burglar's tools for 15 minutes. Better ratings are TL-30, TRTL-30, TRTL-60, and TRTL-60X. This last rating means that the safe can resist attack with tools, torch, and explosives for 60 minutes. There are safecracking methods that UL does not test for that can probably get into the best safe in much less than an hour, but UL ratings and the evaluations of other independent consumer agencies are the best guide at present.

A good move is to back up your safe with its own independent alarm system. Proximity detection is good for this, and vibration, heat, and smoke sensors are a must because some of the best safecracking methods generate plenty of these.

Even if you do not have a safe, a good idea is to get a small, light, cheap one for use as a decoy. Leave it where it is not too hard to find, on casters so that it can be rolled out easily; you do not want it to be opened on the premises. The thief may be so happy to find this that he takes little else. His joy will be quickly gone when, after much labor, he finds the safe empty or filled with a substance that he wishes he had not discovered. If he feels he has the right to break into your home and take whatever he wants, you have the right to give him what you think he deserves.

Ask someone to list the most dangerous items found in the home and you will probably get repetitions of what we have been told over and over by the so-called authorities. We have become so accepting of what "they" say that we often fail to think for ourselves and thereby become easy to deceive and manipulate. Rather than educating the public in the proper handling of hazardous products and substances they are banned by law, resulting in a black market and corruption. Some crime is created by the law itself. Those that steal our property are far less subtle than those that steal our freedom and self-reliance.

One potentially dangerous item found in the home is one that you probably have never been warned about, and that is a home safe that you cannot open quickly when you want to. The safe itself can do no harm; the danger is the impatient and probably armed criminal who demands that you open it. He will not believe that it is empty or that you can't open it, and will be more than willing to persuade you with violence. If precautions as given earlier have been taken, this is an unlikely situation, but the possibility should be considered. Keeping a key or the combination off the premises is advisable for burglary prevention, but is not optimum in the event of an armed robbery. Nothing in the safe is as valuable as your life.

SECURITY ROOMS AND VAULTS

If you are a millionaire with a valuable collection of art, gold bullion, or other such large valuable items, you may wish to build a basement vault. Such a structure is best installed during the construction of a home, but of course it is hard to keep its presence a secret unless you do all the labor yourself, and millionaires prefer to hire their labor. Such are the problems of wealth. Such vaults must be built with all walls, floor, and ceiling of reinforced concrete of eight-inch or greater thickness. Cinder blocks, bricks, or concrete blocks are not nearly as strong. Make sure that there are ventilation holes and that the vault door can be opened easily from inside so that you cannot lock yourself in.

All too often the media bring us a story of vicious criminals or a psychopath invading a home and injuring or killing members of an innocent family. This sort of situation was discussed in Chapters 1 and 4, but some circumstances such as children being at home alone may make an "offensive defense" impossible. For those unwilling or unable to defend themselves aggressively, a security room is a good thing to have; even if you keep a loaded machine gun handy, it is still possible to be taken by surprise or outnumbered. At such times a security room can save lives.

The best room to equip as an emergency refuge is usually the master bedroom, where you will probably be at night when you are most likely to need it. If you live in a multistory home, it should be on the second floor and its windows should not be accessible from a nearby roof, ledge, or fire escape. The room should have the following:

1. A solid, reinforced door and frame with pinned hinges and a high security lock that must be opened with a key from outside the room, but is easily opened without a key from inside.
2. Window gates or heavy shutters that can only be opened from the inside.
3. An alarm panic button.
4. A telephone, preferably on a separate line with an unlisted number, not merely an extension.
5. A battery-powered CB transceiver to summon help if the phone is inoperative and power is cut off.
6. Extra batteries for the transceiver.
7. First-aid equipment.
8. Several flashlights with extra batteries.
9. A fire extinguisher.

10. If the room is on the second floor, a rope ladder for emergency escape.
11. Some means of attracting attention from the window should the alarm not work, such as a loud whistle, airhorn, bullhorn, or firecrackers and a lighter.
12. Front door keys to throw down to the police when they arrive.
13. If you own one, your gun and plenty of ammunition (see Chapter 4, Forewarned Is Forearmed).

If it ever becomes necessary to use this room for defense, never, under any circumstances, unlock the door or leave the room until the police have arrived and searched the premises. The only possible exception is in the event of fire, but use extreme caution and *if your life depends on it,* shoot any intruder on sight. If you have put into practice the advice in earlier chapters, particularly with regard to an alarm system, such dire circumstances should never come about.

OPERATION ID

Many thieves avoid property that is obviously and indelibly marked with information identifying its owner. Operation Identification is a crime prevention program sponsored by most police departments that has been a fairly successful burglary deterrent and has assisted police in both prosecuting criminals, and returning stolen property to its rightful owner. Contact your local police department to find out if this program is available in your area and how to participate if it is.

The police generally provide you with an electric engraving tool with which to mark your property with an identifying code, usually your social security or Federal tax ID number followed by a city code such as "NYC" or "LA." The tool is usually loaned free of charge. Items should be marked twice, once where highly visible and again in a hidden spot. Things such as valuable garments and china that cannot be marked with the engraver should be marked with indelible ink or by sewing a label in an inconspicuous place. Be sure to mark everything likely, or even unlikely to be stolen in both your home and car.

The police will also give you registration cards to be filled out and filed with them, and may also ask you to give them a list of all of your marked property. This is the only aspect of the program that you should have second thoughts about, as although the vast majority of police and police department employees are honest, you would not want such information to fall into the wrong hands. For a burglar who would risk taking marked property, it would be a perfect shopping list.

There are private organizations that provide services similar to Operation ID, but charge a fee. One such organization, headquartered in Virginia, claims to be an organization of police chiefs and uses the name Operation Identification although it is not connected with the Operation ID programs run by many local police departments. According to the Arlington, Virginia Better Business Bureau, this is run by "a for-profit firm." There is no reason at present to believe that it is not reputable, but why pay for a service if it is available free from your local police?

In general, use caution in dealing with any such private services and even with the police if you have reason to believe they are corrupt. Do not give them a list of your property. It is safest to keep such a list of property and relevant personal information to yourself, with the exception of a bona fide insurance company when such information is necessary to obtain coverage. Chapter 10 covers the subject of insurance in depth.

8

LIVING TOGETHER

You would not want to meet Hector on a dark street if you didn't already know him. Everything about this strong-willed, muscular youth left no doubt that he was not to be messed with. At times he did walk the streets at night, which would be very unwise in that neighborhood for anyone else to do. He and a few other local teenagers had joined the Guardian Angels as unpaid warriors in a seemingly hopeless fight against urban savagery and an attempt to show the world that racial stereotypes were unjustified. Where in the '60s youth marched against the bloodshed in Vietnam, some march in the '80s to prevent bloodshed at home. There is hope for us yet!

A common misconception is that the masculine ideal, the macho deity, is unflinching, emotionless, and stoic. Hector made no attempt to hide his tears as he answered the cop's questions. There could be no doubting his sincerity as he answered with a soft passivity so uncharacteristic of the man that gave voice to it.

"OK, kid, you say you found her two hours ago. What made you wait over an hour to call us?" "I was just—just y' know, upset. Findin' her like that. Y' know, she was old, an' I guess she didn't have much time left anyway, but to die like that, it just. . . ." "OK kid. Any of these people see or hear anything unusual in the past few days around here?" "Nobody said anything to me till today, when I got home. Why don' you ask them?" "Look, angel boy, I'll run you in for interfering with police work if you don't answer me straight and fast, and I'll shove that red beret down your throat for good measure. Now what did you hear today and who did you hear it from?" "From Cynthia, who lives in 4D. She met me outside the building an' said that she wouldn't answer the doorbell, and that there was a bad smell comin' out. I thought it was garbage an' that she was just gettin' more hard of hearin', but when I came up . . . I came up an' the door wasn't locked. I pushed it open an' there she was." "Yeah, it looks like a push-in job. Her purse is empty and the apartment was turned inside out. They probably only meant to knock her out, but hit her too hard. You got any ideas who does this crap around here?" "I wish I did, oh God I wish I did, oh God, God . . ." "Hey, stop acting like a wimp and listen to me! Who was she? What was her name?" "Maria Sanchez. She was my grandmother."

PROBLEMS OF APARTMENT BUILDING SECURITY

Most of what has already been discussed applies to both private and multiple-family dwellings. The nature of multiple dwellings (which most often are apartment buildings that house hundreds of people) creates security problems not found in private residences with a single family or three at most. These vulnerabilities are well known to burglars and their ilk and are best remedied by vigilance and cooperative action by the tenants.

If we examine the situation of an apartment building with 150 tenants in a relatively low-crime area, assuming a 1 percent criminal population, (neglecting "victimless" crime, which might make it 50%), the mathematical probability is that there are one or two serious criminals living in the building and only a 22 percent chance that there are none. If we have 10 percent of the tenant population with predominantly antisocial traits, not overtly criminal but the unfriendly, gossipy, or backstabbing sort, we have another roughly fifteen tenants who will covertly work against the crime prevention efforts of others or at least refuse to cooperate. This is in a "good" neighborhood.

A 150-tenant building in a low income, high-crime area is far worse off. Assuming a 5 percent criminal and 20 percent antisocial population density, there are roughly seven or eight serious criminals with only a 0.046 percent chance of there being none in the building. There are also approximately thirty other undesirable individuals. Pity the roughly 112 decent human beings forced by social and economic circumstances to live with these in their midst!

The tenant of a multiple-dwelling building, particularly a large one, cannot assume that all of his neighbors are trustworthy. It is almost certain that some are not. Your enemies being among you is probably the worst but is far from the only security problem in such a building.

Two groups of people are highly vulnerable to crime and likely to be careless when it comes to security precautions—the very young and the very old, particularly the latter as they often lack the necessary care and supervision. Because of disability, apathy, and senility, some will leave doors open or unthinkingly admit strangers to the building no matter how many times you ask them to use caution. They often wind up the victims of their own carelessness.

Unless your building is a modern one built with some consideration for security, you probably have such structural vulnerabilities as old wooden or paneled doors, alleys from which fire escapes are accessible, dumbwaiter shafts, and a roof accessible from adjoining buildings. Even fairly modern buildings are very easy for an intruder to enter, as lobby door latches are usually poor security that can easily be defeated with a plastic strip or a good strong shove on the door (see Figure 8-1). If that fails, one can always push a few intercom buttons—it will only take a few tries in most cases before some fool will buzz open the door without even asking who is there. An alternative is a basement or garage door that may be left unlocked. An

Figure 8-1. Lobby door insecurity. The cheap cylinder has been drilled and forced, and the wind prevents the door from closing anyway.

unusually conscientious landlord may have originally provided very good deadlatch locks with protective hardware and a video intercom system at all entrances, only to have them repeatedly rendered useless by vandalism. The superintendent knows that if he fixes them again, they will not last a week. The local burglars know this and come and go as they please. Not only apartments, but storage rooms, workrooms, garages, and mailboxes provide loot and are usually very accessible. Of course if a burglar can easily enter a building, more violent types can too, and elevators, stairways, and halls provide criminals with more concealment than the streets for the exercise of their savagery. Antiburglary measures will combat more serious crime as well.

Upper-income, high-rent apartment buildings are often equipped with closed-circuit TV, built-in alarm systems, well-maintained locks, and intercoms. They usually have a doorman and a garage attendant, and are relatively free of the sort of crime that plagues less secure buildings. The obvious affluence of the tenants here attracts another type of criminal, the highly skilled professional burglar. In spite of all routine precautions, the pros do well here. Residents of such buildings should never rely on management-provided security alone.

Landlords who live elsewhere may think of their apartment building in the abstract way one thinks of the inner functioning of a company in which they own stock. Their goal is to maximize profits. Tenants may appear to work against this, and therefore are considered enemies. Many landlords are beset with harassment from a few antisocial tenants and as a result develop an antitenant bias that victimizes the majority of tenants who seek to work with the landlord for their mutual benefit. The fact is that an investment in building security has high returns in terms of maintaining property value by preventing vandalism. Many management people are too shortsighted or biased to see this, and the tenants are left to act on their own. Some mention was made in Chapter 2 of dealing with an uncooperative landlord on an individual basis, but action taken by a well-organized tenants' group is more likely to be effective.

TENANT ORGANIZATION AND COMMUNICATION

Although any multiple-dwelling building may have its share of criminals and other undesirables, fortunately it can also have its share of community activists with leadership ability. If you are concerned enough to be reading this book, you may be one such person. If you feel that someone else is better qualifed as an organizer, present the idea to them and offer your support. You will be one of the most valuable members of your tenants' association by virtue of having the information in this book, and your desire to motivate others for your mutual benefit. Your initial goal will be to get better acquainted with your neighbors. Through communication, cooperation, education, and action a powerful group will be established whose goal is to make your communal home safe and secure. You will also be able to identify the seriously antisocial and criminal individuals among you and if necessary take action against them. It is certain that they act against the common good and endanger you all.

Need you fear eviction, refusal to renew your lease, or landlord harassment because of your tenant activism? In general, no. In New York State for instance, the landlord is forbidden by law to take such measures, except if the multiple dwelling has fewer than four units and is owner-occupied. In fact, should any action be attempted, the law *presumes* that the action is a consequence of the tenant's activity, and the burden of proof for legitimate grounds for his action lies with the landlord. State and local laws vary and you should check yours, but in general, it is your legally protected right to organize or join a tenants' association and take justifiable legal actions against the landlord when more amicable measures fail.

A tenants' association should have access to a photocopy machine so that information can be quickly prepared for distribution to all tenants. Do not rely on word of mouth communication or a notice on a bulletin board. When distributing leaflets, make sure to push them all the way under the door so that someone else cannot remove and discard them. At times you may wish to withhold information from uncooperative tenants, and communicating this way allows you to do so. If some people do not speak English, be sure that the information is translated into their language as well. The active participation of everyone in the building is nearly impossible to achieve, but try for the most you can get and always maintain a friendly approach to avoid alienating anyone. You can expect opposition and even dirty tricks from antisocial tenants and the landlord, but once it is clear that you are a strong group that will not be pushed around, you will get respect.

Almost all neighborhoods have a local politician that is a tenant's advocate, and it is important that this person be contacted when you are forming your group. Priceless information and advice will be made available to you, and if you convince this person that you have considerable influence in the community, he or she may prove to be a powerful ally whenever the going gets rough. Existing neighborhood or block associations can give useful advice as well, and affiliating your group with theirs may be beneficial.

The first step for all tenants is to improve the security of their apartments, then to work as a group on building security. Many local police precincts provide crime prevention education services and will gladly send an officer to speak with your group. Precinct community council meetings or similar events give an opportunity for dialogue that is valuable in establishing a good reputation for your organization and for making specific problems known. Special services for crime victims and free security evaluations for your building may be available from the police, although the information given here enables you to do a good security evaluation of your own. If available, take the police service, as they might find something you overlooked.

If some issue other than crime is of greater concern, it may be used as a basis of tenant organizing. The initial approach to people should emphasize that whatever the common goal or enemy, there is strength in numbers. Even if you enjoy a low crime rate at present, the situation could change for the worse if you do not prevent it. Organize and be vigilant; vigilance is the eternal price of liberty!

SPECIFIC MEASURES

Laws concerning apartment building management, security, and tenants' rights vary among localities. In many areas, the management of multiple-dwelling residences is required by law to provide and maintain self-closing and self-locking doors at all entrances, intercom systems, apartment door locks, peephole viewers, and hallway corner and elevator viewing mirrors. They are usually required to permit tenants to provide their own doorman or attendant services when not provided by management, and to permit tenants to install their own locks and other security devices in addition to those already present. If a tenant falls victim to a crime in his building or apartment and can show that the criminal took advantage of the landlord's negligence, he may have grounds for legal action to force the landlord to pay for damages.

All tenants should become thoroughly familiar with their building inside and outside, including the areas they do not normally use. In this way, possible danger spots can become known and people can offer ideas for improving security. All the tenants must realize that it is their responsibility to see that the superintendent or landlord quickly repairs defective locks, replaces burned out light bulbs, and promptly attends to other routine safety related matters.

Your tenants' group will come up with recommended building security improvements that inevitably will require money and labor. This is where tenant-landlord conflict is most probable. Friendly persuasion is the optimum path; also effective is obtaining information and statistics from local authorities, real estate agents, and insurance companies that tangibly demonstrate that property values are enhanced by good building security. Tax credits and deductions may be available for the costs incurred by the landlord; find out if this is the case. If so, he probably already knows it, but it enhances your position to point out such facts, and makes him grateful if they were not already known to him. Such information could be worth far more than the price of a few new locks. If any insecure buildings in your area have suffered from vandalism, show the landlord photographs of them contrasted with better protected buildings. Speak in terms of his interests as well as your own and you may hit a responsive chord. Should this approach fail, perhaps the tenants could collectively bear the cost of the needed work and, if possible, offer to do it. Continued intransigence makes it necessary to look up the law or consult your politician friend for advice. Hiring a lawyer is expensive and does not guarantee any better results than you can achieve independently.

Tenants can take action through state and local government agencies at little or no cost if the landlord may be in violation of the law or evading obligations. Be aware though that a landlord may be entitled to a rent increase should substantial improvements be made that are beyond those required by law. Some landlords even get away with calling necessary repairs "major capital improvements" and get rent increases on that basis. If the landlord is stubbornly uncooperative, relatively small matters like installing a better deadlatch rimlock on a side or rear entrance door can be done once you have made the landlord aware of your concern for security. If the superintendent is friendly, you may pay him to do it, or you can all chip in for the lock and let a skilled tenant install it. A locksmith provided with the key can arrange the cylinder pins so that the old keys fit the new lock. Should your landlord refuse to permit such improvements, point out to him the high cost of vandalism in buildings with inadequate security. You may be able legally to force him to accept the improvements; otherwise he may have to learn the hard way when a crime occurs for which he is indirectly responsible.

In general, the beefing up of building security will require the application of all the information given in earlier chapters on the criminal's methods, doors, locks, keys, windows, and "other ways."

Go over this with your tenants' association and methodically see to it that as much is done as possible. This will not happen overnight and you may run into problems, but with determination and mutual support you can move mountains. Here are a few things that must be done without fail:

1. All building doors must be very physically strong and equipped with hinge springs or other devices to make sure they close and lock automatically, even in a high wind. High-quality deadlatch locks must be used for automatic locking; preferred is the type with the strike constructed so as to shield the latch from attempts to open it with a shim. Lobby door locks that can be buzzed open with a button from the apartments are particularly vulnerable; close-fitting guard strips and plates must be installed to make shimming or prying impossible.

2. Locks and cylinders on lobby and other doors in constant use wear out much sooner than those on apartment doors. They should be lubricated at least once a month and replaced at the first sign of trouble. If a key can be removed from the slot before it is turned back to the vertical position, the cylinder is worn out and should be replaced immediately.

3. Locks must be located on the outermost doors wherever possible so that someone cannot work at defeating them sheltered from view from the street.

4. Every evening the tenant patrol, superintendent, or other responsible individual should carefully check the locks on all building entrances to be sure they are working properly. Defects must be immediately repaired. Evidence of sabotage should be reported to the police, and all persons should be on the alert for intruders at such times.

5. All tenants have building keys, so all must be educated in key security. Although it will require some expense (not much when shared by all), the lock cylinders should be changed every year or so and new keys should be distributed to all tenants. If an undesirable tenant moves out or is evicted, the cylinders and keys should be changed immediately.

6. Fire stairs should have one-way doors that can only be entered on the residence floors and exited from the roof and the ground floor. All exits should be equipped with alarm locks.

7. All windows in the lobby, basement, hallways, garage, storage rooms, workrooms, boiler room, and so on should be well-secured, preferably with steel bars. There may not be much to steal in the boiler room, but once a burglar is inside, he has access to the rest of the building.

8. All doors to the roof should have alarm locks.

9. Fire escape ladders must be made inaccessible from the ground.

10. Garages, basement storage rooms, isolated laundry rooms, and workrooms must be well-secured. It should be impossible to enter the garage without a key or garage door opener, and only those tenants having a rented space should have one. Garage doors should automatically close within twenty seconds if not closed sooner. Only the superintendent should have a storage room key, and it should never leave his hands.

11. A good intercom, preferably video, should be maintained and all tenants must use it at all times to avoid admitting strangers to the building.

12. Tenants must be taught to be alert for anyone loitering by the outer building doors and never to open such a door when it may enable a stranger to enter.

13. All tenants must be instructed to call the police and then the tenant patrol should anyone or anything suspicious be seen or heard.

14. The building must be well illuminated at all times inside and at all hours of darkness outside.

It is very important not to permit strangers to enter the building by either carelessly buzzing in anyone who rings the bell or by holding the door open for someone when entering or leaving. It is difficult to enforce this, as inevitably there will be tenants who through disability, carelessness, or worse will not cooperate. You should distribute leaflets that clearly and carefully explain the need for caution, and explain the operation of the intercom and door opening button for the benefit of any who may not know this. Also tell tenants to reply to an entry attempt by a stranger with the statement "I'm calling the police."

Regular security tests of the intercom/door-opening system should be conducted to detect, educate, and if necessary take action against careless or uncooperative tenants. Choose a time when most people are home, such as 7 PM on a weekday, but do not always do the test at the same time. Have a prepared list of all the apartments, and ring each bell one at a time and wait for a response. If tenants ask "Who is it?" (assuming there is no closed-circuit TV) you may say "delivery" as a ruse to get them to open the door, which is what a criminal might do. If they refuse to open it without being certain of your identity, explain that it was a test, thank them for their caution, and put a check next to their apartment number on your list. Should there be no response, put an "N" by their number on the list and be sure to test them again at a later time until a response is obtained. If they open the door without using the intercom, or as a result of your trick, do the following:

Have leaflets prepared worded as follows: "Dear neighbor, at (time) on (date), a building security test was conducted by your tenants' association. When your bell was rung, someone in your apartment opened the entrance door either without using the intercom or without getting a satisfactory answer as to the caller's identity. LETTING STRANGERS INTO THE BUILDING ENDANGERS YOU, YOUR FAMILY, YOUR PROPERTY, AND YOUR NEIGHBORS. Please don't make this mistake again. Your tenants' association is working without pay to make our building a safer and better place to live. Please cooperate. This test will be repeated in the future."

Distribute these to those who have an "X" next to their apartment number on your list. A week or so later, repeat the test. Those who score an "X" for the second time should be visited in person, preferably by a neighbor who knows them and can impress them with the importance of the matter.

Unfortunately, some people are like Pavlov's dogs and will buzz open the door unthinkingly no matter what they're told. It would seem as if they want to be victimized. Unfortunately, they endanger everyone, not just themselves. You must act in the interests of the majority and not allow the negligent or malicious few to subvert security; your strong action may literally prevent a murder.

If several personal appeals are ineffective, a petition may be signed by all other tenants and sent to the offender by certified mail. This should warn them that if their careless disregard for the safety of others continues, the landlord will be petitioned to evict them. Keep a copy of this petition for use should it be necessary to prove that the person had been warned. In the petition to the offending tenant, state specifically what the objectionable conduct is, such as buzzing the door open for strangers or holding it open for them when entering or leaving the building. Should this measure not be effective, resort to an eviction petition. Prior to this, it would be wise to discuss the situation with the landlord, who more likely than not will want to cooperate.

When all other measures have failed, some tenant groups have been known to open the intercom bell panel by the door and cut or disconnect the wire to the offenders' buttons whenever they fail a security test. This however should not be done as it is likely to be disapproved of by the landlord and may be illegal.

After several months of security tests you will know those tenants who are always careful, and it will not be necessary to keep testing them. Frequent tests of those tenants who scored an "X" in the past should continue, and new tenants should be tested several times until you are sure that they are always careful. Remember that cliche about vigilance.

The presence of a medical, dental, or other professional office in a residential building can create a dangerous situation unless the office has a separate outside entrance that is well-secured and used by all visiting patients. In the office lease, landlords should specifically forbid the use of any other entrance or exit by such visitors. If no separate entrance exists, the office receptionist should use the intercom at all times to avoid admitting strangers, and watch patients leave to make sure they do not loiter in the building. Some doctors willingly cooperate, others arrogantly refuse. Take appropriate action against them, as you would for any other uncooperative tenant. Unlike residential apartments, professional suites are rented under a commercial lease that may lack the legal protections of a residence, such as a guarantee of renewal under the same terms as the old lease. Knowing the relevant laws, tenants on good terms with a landlord can force compliance when such people are uncooperative.

Someone attempting illicit entry may wait until the building's garage door is open and enter undetected before it closes. A garage door should never be used for entry or exit without a vehicle. All garage users should be notified of this in writing, as well as the following:

* When entering the garage, do not park immediately but wait in your car, watching the door until it is fully closed. If a stranger enters, sound your horn as an alarm and open the door again with your remote control. If he does not leave, back out of the garage and get the police. Never get out of your car in the garage when a stranger is present. If you cannot open the door from inside, keep sounding the horn in an SOS or on-off pattern until he leaves, or help arrives. If you are in serious danger and have no other weapon, use your car.

* When leaving the garage, look around outside and wait for the door to close before driving away. If an intruder enters, immediately get the police and let them into the garage.

* Garage door openers and keys should never be left anywhere inside a vehicle or garage, even for a short time.

If a tenant is known to be a serious criminal or to be endangering others by allowing the criminal use of their premises, the police should be notified. If arrest results, eviction may be more easy to accomplish. Eviction petitions are a last resort means of eliminating the threat from within, and in such cases, landlords are usually cooperative (it can profit them) but politicians are not (it can make them look bad, and they always leave the dirty work to subordinates). Criminals will often get paranoid in an atmosphere of anti-crime militancy and gladly leave on their own. Goodbye and good riddance.

There will be times when workmen will be present in the building or will be entering and leaving it frequently. An understanding should be reached between your tenants' group and the landlord or resident superintendent on their supervision in relation to security. Under no circumstances should they be permitted even brief possession of any building key; the superintendent, doorman, or other responsible individual should admit them when necessary and make sure that all of them have left at the proper time. Tenant patrol persons should be alert for any suspicious conduct, vandalism, sabotaged locks, or evidence of criminal activity, and report any findings to the police or the landlord as is appropriate.

If your superintendent and other building employees are honest, reliable people, they are an asset to you and it is worthwhile to maintain a very good relationship with them. The tenants' association can serve as an impartial mediator in any dispute between them and a tenant, and can act in their behalf should any problem arise that endangers their jobs. Should a building employee be incompetent or untrustworthy, a tenants' group can be effective in getting them replaced. Be sure to have plenty of hard evidence and testimony available to back up your complaints before contacting the landlord or police.

Where the funds are available, surveillance CCTV systems, hallway intercoms, and alarm boxes may be installed and monitored by a hired doorman or security guard. Such protection should be manned twenty-four hours a day. The same caution in hiring should be observed as when selecting a central station alarm or other guard service.

PATROLS

Citizen patrols, whether community-wide or limited to one building, vary greatly in their acceptance by the local authorities. In some areas they are encouraged and supported, given training, walkie-talkies for communication with the precinct, and gratitude for their concern. In other areas they are treated as if they were criminals themselves. As much as possible you should communicate and work cooperatively with the police, but do not be discouraged if the police oppose you and fail to provide the protection you require. Here is where a politically active ally can be very helpful.

Citizen patrol groups incur expenses that may initially have to be paid for by donations from local residents and businesses. Once a group has proven itself to be ethical and effective, grants from state and local budgets may be available, especially if it is working in cooperation with the police and local government.

As with a tenants' organization, existing groups such as block-watchers, street patrols, and other tenants' groups can provide valuable information. If the police are supportive, take advantage of all the help they offer; training and the proper equipment can make the difference between an effective, well-organized patrol and a few vigilantes who wind up in a shootout with no way to summon help.

Be sure that there are enough active members to assure regular, preferably two-person patrols. Do not operate in a predictable pattern; one evening you may patrol from 9 PM to midnight, another from 8 PM to 10 PM, and another from 3 AM to 7 AM. These last hours are best put in on weekends, when garage and storage room burglars are most active. Nothing can make an impression on such a lowlife like finding himself looking up the barrel of a 12-gauge shotgun. Your command to stay put until the cops come will be obeyed implicitly, and the word will quickly be out on the street that yours is one building or block to stay away from. Nightsticks may be sufficient for ordinary patrolling, but 5 AM in a basement or garage calls for "heavy artillery." Of course, armed patrols bring up questions of safety and legality that should be carefully considered before undertaking such action. It is not for the timid or trigger happy.

Less militant blockwatcher groups are effective as extended eyes and ears for the police. Once they have been successful in causing a few arrests, crime in the area will most likely decrease noticeably. Blockwatchers operate with no risk to themselves, as the police give volunteers an identifying code so that they can call in reports anonymously.

Organizing a blockwatch or buildingwatch is not difficult, especially if a tenants' association or similar group already exists:

1. Call a meeting at a convenient time to be held in someone's home or a public place such as the building lobby, a school, church, or the police station. Distribute leaflets to all tenants or residents of the

block well in advance, telling them of the meeting, its purpose, and importance. Mention that a police representative will be present to discuss available crime prevention services such as free home security surveys and Operation Identification. Emphasize that the meeting is free and that attendees will not be under any obligation.

2. Invite the local precinct Crime Prevention or Community Relations Officer to attend to explain the available police services and give advice and encouragement regarding the block or buildingwatch operation.

3. Begin with individuals briefly presenting their experiences and concerns to which the police representative can respond. Show how such incidents can be prevented when the people of a community work together for mutual protection. At the end of the meeting take the names, addresses, and phone numbers of those who are willing to participate.

4. Prepare a list of all members of your block or buildingwatch group, with each name, address, and phone number. At the top of the list, give specific instructions for the reporting of suspicious activities to the police. Some areas have an emergency phone number such as 911 for quickly reaching the police, but it may be better to use a direct call to the precinct (the police representative can provide you with several numbers). The first action should be to call the police, then call the number for the person whose premises are involved. Except possibly for a life and death situation, no action should be taken beyond making these calls and continuing to observe what is happening, in order to inform the police when they arrive.

When the list is complete, make copies and distribute them to (and only to) participants.

5. If possible, arrange to provide all members with a large, loud whistle to be carried at all times and used to summon aid from the street in an emergency. The whistle should be blown urgently only when necessary, and anyone hearing it should call the police immediately.

6. The group should meet at least once a year to exchange information on current neighborhood crime problems, recruit new members, and update the list of members. Make sure that a police representative is present at these meetings.

7. Should a member become a crime victim, call a special meeting. Discuss what happened and what measures should be taken to prevent such an incident from happening again. The group should provide the victim with what the "criminal justice system" does not; compassion, assistance, and support. Offer to accompany them to the police station, the District Attorney's office, and court, and with any

other necessary procedures. If they are indigent, take up a collection to defray their expenses. In short, treat them as you would wish to be treated, and in so doing show them that there are still some decent people left in the world.

Whether you are a member of a citizens' patrol, a blockwatcher, or just a concerned human being seeking to make this a better world, this is when and how to call the police:

1. If you see any suspicious-looking loiterers or even ordinary-looking but unknown people lingering anywhere for no apparent reason.
2. If you see anyone looking into or entering a neighbor's home or apartment.
3. If you see anyone looking into or tampering with parked cars or "fixing " a car that you know belongs to someone else.
4. If you hear an alarm, breaking glass, screams, shots, or dogs barking abnormally. Never ignore an alarm, even if you think it may be false.
5. If you see "workmen" or "movers" working by or on a house or apartment when the owners are not at home.
6. If you see any door-to-door salesman or solicitor trying to open a door or going into an alley or backyard.
7. If you see anyone suspiciously carrying unwrapped property or running while carrying property.
8. If you see anyone using a portable communications device in a suspicious manner.
9. If you see a stranger park his or her car nearby (in a residential area) and walk far away from the place he or she parked, for no apparent reason.
10. If you see a stranger dropped off from a vehicle behaving in any way differently than an honest visitor would.
11. If you observe a vehicle (other than a patrol) cruising slowly back and forth or any vehicle or pedestrian returning at intervals and seemingly observing the area.
12. If you see any stranger in a parked car in a garage, driveway, or in front of a home, especially at an unusual hour or when the residents are not home.
13. If you see any evidence of a crime such as a door or window broken open or bloodstains.

Patrol and blockwatcher groups have the advantage of quicker and higher priority communications with the police via walkie talkie, CB, or telephone. If you feel the police should be called, do not delay for fear of causing a false alarm. If you bring about the arrest of one criminal or prevent one crime, a few false alarms are worth it and the police will be grateful for your help.

The police themselves can tell you the best way to contact them in your area. Nonemergency conditions should be reported to the local precinct. Many localities have an emergency number such as 911 or 0 (operator) for quick response, and this generally is to be used if you have any reason to believe there is a crime in progress or one has occurred very recently. When you get through, first state the nature of the emergency: "BURGLARY IN PROGRESS," "HOLDUP IN PROGRESS," or whatever. Give the exact location of the incident and if you can, describe the perpetrators and/or vehicles involved. If they have fled, tell the police where they were going when you last saw them. In an apartment building, be sure to open the entrance door for the police when they arrive so that they are not delayed. Tell them what apartment to go to and give them any essential information such as to cover a rear or side door to prevent a criminal's escape. You know the premises better than they do, and a few helpful words can make a difference. If you cannot be in the lobby when they arrive, use the intercom.

Police forces, particularly in densely populated areas, are often overburdened and should not be bothered with minor complaints that they may decide to ignore anyway. Nonviolent domestic quarrels, kids making noise in the street, people making love in the bushes, and dozens of other trivial violations may seem objectionable to you but should not take police time away from more serious matters. If you suspect a real crime, call them without delay, but leave chasing stray dogs to the ASPCA.

People who would like to devote their free time to the betterment of society can do so by forming court monitoring groups. These groups attend criminal proceedings to protest against the overly lenient treatment of offenders and any disregard shown by the court for the victim's rights. The judges and prosecutors who fail to protect the public become known, and through public pressure, are forced to either better serve the people who pay their salaries or face the loss of their positions.

Citizens should not take the law into their own hands, as law enforcement is a dangerous enough job when you are trained for it.

Sometimes, however, when the established powers are unable or unwilling to provide adequate protection, some action must be taken. The Wild West was civilized only because the decent people had more guns than the criminals. Hi-yo Silver, Away!

THE LESSER OF EVILS

In a totalitarian police state, there are few freedoms, but little crime. In an anarchy, there is great freedom but plenty of crime. While a maximum of freedom is desirable, a balance must be struck between the rights of the honest citizen and those of the malefactor; an obsession with the rights of the criminal, with little regard for protecting the public from him, only encourages criminality and the decay of society.

There are some laws on the books, often enforced selectively, sporadically, or not at all, that we would be better off without. These are the widely flaunted laws of prohibition, the violation of which harms no one except possibly the violator. These laws encourage corruption and undermine public respect for the law in general; they actually may create rather than prevent crime, as many of the real evils associated with the prohibited activities are due to their very illegality.

As balance must exist between the rights of the criminal and public protection, so must it exist between the fire protection laws and the need for security from intrusion. Just as the victimless crime laws were perhaps better suited for the time when they were written than to the present, so may some of the building codes be in need of revision in light of the present crime rate. The fire codes vary between localities; their purpose—a very good one—is, among other things, to assure that residential structures allow a way of escape in the event of fire blocking the usual means of exit. Perhaps in the urban America of 1950 more lives were saved by apartment building fire escapes than were lost to criminals who used them to gain access, but is this still true today? Is more property now lost to fire or to theft? This of course varies between urban, suburban, and rural settings, but any residential locality that has experienced a marked increase in crime in recent years would do well to consider if its building codes impose excessive vulnerabilities and if these should be remedied.

In some cases the law is truly the burglar's friend. The building codes of one particular city—which to prevent an influx of criminals will not be named here—specifically limits the thickness and type of glass or plastic that may be used for skylights. Another section of the codes in this thieves' paradise places certain requirements on apartment building balconies that may be used as a means of egress. Each balcony must serve at least two apartments, and the doors between apartment and balcony MUST have large glass panels free of obstructions that can hinder breaking the glass. The inside locks on these doors are REQUIRED to be easily openable from outside after the glass is broken; combination, key, or tool-operated locks are forbidden. Should the need arise, other tenants or firemen can enter easily. So can burglars or an arsonist.

Codes governing multiple-dwelling buildings frequently require fire escapes, the means of access used in the majority of apartment burglaries where entrance is through a window (see Figure 8-2). These codes prohibit the most effective means of securing easily reached windows, such as permanent bars and key-operated locked gates. They may forbid the use of double cylinder secondary door locks due to the possibility of delay in escape caused by misplacing the inside key. They often require fire exits in garages and other public areas of multiple-dwelling structures. Inspectors make regular rounds to assure that such exits are in working order; landlords may be fined if they are not. Does anyone come around, other than the criminal, to see if these buildings are intruder-proof? No, and the crime victim pays a heavy price for the negligence of both the law and the landlord.

One of the first considerations of a burglar is a means of quick escape should he be discovered in the act by a resident or the police. He does not want to be caught, and usually prefers to flee than to fight. Given a choice between two similar apartment building garages, one with a rear fire exit and one without, he will hit the one with the exit, even if he cannot use it as his means of entrance. Such exits are required by law to be easily openable from the inside, and provide an excellent route for escape. Poorly secured, they are an easy way for him to get in as well. Are such doors really desirable? The decision must be based on recent local statistics of losses due to crime as well as fire; to combat one danger by creating a greater one is self-defeating.

In spite of all precautions, lives and property will be lost to both fire and crime. The criteria for the formulation of building codes should be the minimization of risk from both, not from fire alone. Present laws should be reviewed and revised accordingly, and appropiate alterations made in existing buildings wherever possible.

Figure 8-2. The city burglar's dream come true.

9

CAREFREE VACATIONS

Burglars love vacations. Long holiday weekends are their favorite times of the year. While others are planning to leave the routine and get away from it all, they plan a routine to enter and get it all. Some thieves plan with amazing brilliance and ingenuity.

Cal had a knack for numbers, and when he first laid hands on a computer terminal in high school a love affair began to which all else in his life took second place. He really didn't know the difference between AC and DC, and he thought a RAM was an animal in the zoo. He had no interest in becoming an engineer, having been convinced by his father, an electronics designer, that they were not treated by employers like the professionals they are. Before graduating high school he had decided he would do better than any engineer or engineer's boss without even going through college.

Cal was expert in BASIC programming and had his own home computer, but found it limited and not able to access certain data sources without more equipment and information. At the age of eighteen, Cal had spent more time among machines than with friends, and having been well indoctrinated with materialistic values and moral hypocrisy, he enrolled in the Kludge School of Computer Sciences with unspoken intentions never suspected by his parents or instructors. He avoided most courses in electronics and information theory; his burning desire was for practical hands-on training. At the end of two years he was fluent in all computer languages and could almost talk to a computer directly through a modem by vocalizing beeps of varying pitch.

Cal had an elaborate home system by this time, with all sorts of disc drives, practically infinite memory, and some unusual software useful in decoding scrambled data. He turned down a few opportunities but grabbed the chance to work for a large airline. The job paid less than even a technician position, but Cal was patient and through manipulation, neat appearance, and competence worked his way up to a supervisory position within two years.

Cal now resides in a minimum security correctional facility, a euphemism for a jail for white collar criminals. The popular term is a country club prison, and that is an apt description as he lives better than many people with a 9 to 5 job who pay taxes and support a family. After playing his game for eight years he half expected to be caught, as it was impossible for him to stay one step ahead of all the new technology. He planned for his downfall well though, and when he gets out in a year or two he will have hidden assets such as buried gold and Swiss bank accounts to live well on for the rest of his life.

Cal's game was to take a little basic information from the airline's passenger lists, often no more than name, address, and a credit card number, and with a little computer finesse, obtain just about any information he wanted to know. When he found an individual of suitable affluence and apparent vulnerability he would forward the data to professional burglars who would pay as much as a hundred dollars for a two minute phone call. Cal misses the excitement and ready cash, but looks forward to a contented retirement at the age of 34. If he gets bored, he is sure he can find some politician or agency that would be glad to hire him.

The subject of security while away from home for an extended period has been held off until now for good reason, because most of what you should know and do has already been discussed and only a few additional points need to be made. The topics "Privacy and a Low Profile" in Chapter 1, "Alarms" (Chapters 5 & 6) and "Stash It Away" (Chapter 7) are particularly relevant.

A tremendous amount of bad advice has been publicized again and again on this subject to the delight and enrichment of many criminals. Most of it seems sensible and protects you in one way, but leaves you vulnerable in another.

BAD ADVICE

When you go on vacation, make sure to have all deliveries stopped, tell the phone company to disconnect your line until you return, and have the post office hold your mail for you. Leave a light on and a radio playing near the door. Tell all your friends and neighbors so that they can watch your house or apartment, and leave your small valuables with them for safekeeping. Put tags with your name and address on all your luggage to prevent loss. If you can, hire a house or apartment sitter or let a friend move in for the time you will be away, or at least close up all the blinds, curtains, and shutters so nobody can look in. It bears repeating: *This is all bad advice.*

If you have good security in your home or apartment, having someone else, even someone trustworthy, stay in it during your absence is a greater risk than leaving it unoccupied. Will they always use all locks and the alarm when out? Will they admit any people who are less than trustworthy? Will they at some future time divulge information about your home to others? The risk is considerable. If you have pets or plants that require care, leave them with a trusted friend rather than have someone come in to care for them. Small valuables belong in a safe deposit box; they may be less safe in a friend's home than in your own.

SECRECY AND ANONYMITY

Your aim is secrecy at home and anonymity when away. Follow bad advice and dozens of people, perhaps hundreds, can learn that you are away from home. They work for the phone company, post office, service companies, and delivery agencies. They work for the airline or any other transportation service that handles your luggage. They work for, or may be the guests of, any motel or hotel you stay in. They are your neighbors and business associates to whom you boasted of your upcoming vacation. They are people who casually walk or drive by and see that your home looks shut up or has one light on night and day. If 100 people know, chances are good that one criminal knows. Don't get paranoid, get wise.

There are other subtle ways of tipping off burglars as to your absence, such as being seen loading luggage into your trunk or leaving a loaded camper or van in your driveway or in front of the house before leaving. A garage that cannot be seen into from the outside is the place for such things. Every action you take should be considered as to the possible risk incurred. This might seem paranoid and unnecessary, but before long you will develop habits as easy to follow as those you are presently accustomed to, but much safer.

You should single out your most trusted neighbor for a mutual security service agreement. When one of you is away for any length of time, the other will do all that is possible to make the unoccupied house or apartment appear to be in daily use. This means receiving all deliveries (if you both get milk he can stop his delivery for the time and take yours), emptying your mailbox daily, removing any papers or leaflets left by your door daily, mowing and watering the lawn, shoveling snow from the sidewalk, using your garbage cans, parking in your driveway, and so on. He also should be the one given the phone number where you may be reached if any emergency arises that he cannot handle. One benefit of tenant and block associations is that you get to know people who are trustworthy enough for such an arrangement.

If you must temporarily suspend any service or delivery, do so without explanation or with a misleading one such as "I'm on a diet and won't be drinking milk for a week" or "I don't care for that newspaper anymore." A post office box or a mail drop can be used both to hold mail and as an address when registering with a hotel, travel agency, or airline. A business address may also be used in place of your home address on baggage tags to assure privacy, and travel services are available that provide coded luggage labels. Try to avoid keeping papers with your name, address, or any other personal information in your luggage, hotel room, or where it can be casually observed.

The phone should never be disconnected, but left off the hook if your absence will be a short one. If you will be away for several days or more, you might consider an answering service that will forward your calls, or preferably (for greatest privacy) an answering machine that you can call from time to time and, upon your signal, receive the recorded messages. You can then return the calls without the callers knowing that you are not at home.

Leaving a car parked in your driveway twenty-four hours a day seven days a week is a sure sign that you are away. If you must put a car in storage, do not say for how long it will be, or say "I may be back for it in a few days but I can't be sure when." This will keep them honest with your car as well as your home; you should not indicate that you will be out of town, but that the car is being stored for convenience. The best course of action is to leave the car with a trusted friend or relative if you cannot secure it in your home garage.

Children must be taught the reasons for secrecy regarding periods spent away from home and not merely ordered not to say anything about it to anyone. If you feel a child is prone to spill the beans, if possible keep the upcoming vacation a secret from him or her until the last minute.

Frequent checks of all locks and the alarm system should be made, and no time is better than before a vacation or other absence from home. Be very thorough, and never leave until everything is in proper working order. Before leaving double check your security; be sure that the timers for lights, radio, and TV are properly set and that the appliances themselves are turned on, all unnecessary appliances such as the stove and hot water heater are off, all windows are closed and locked, and all other locks are in use. In winter your heating system should be set to maintain a temperature of about 60°F. Window blinds, automatic outdoor lighting, automatic lawn sprinklers, and so on should be left as normal. When you leave activate your alarm system, lock the doors, and be on your way. If you have followed all this advice, your property is as safe as anything can be in this world.

A HOME AWAY FROM HOME

Structures that are unoccupied much of the time, such as a vacation home, second residence, or a farm or ranch building pose considerable security problems. As crime has become a popular vocation in urban and suburban areas, it has grown in our rural lands as well, although not to the same extent as in the cities. In our more innocent past, cabins in the remote woodlands were left unlocked and stocked with emergency rations for any lost or injured person in need of food and shelter. Today it wouldn't be long before such a place would be taken over by squatters, vandalized, cleaned out, or totally destroyed. Our civilization has not kept pace with its technology.

All the security measures discussed in previous chapters apply to homes and buildings in rural areas as well, but some are rendered less effective by the fact that the thief may have plenty of time to work, with little chance of being seen or heard even if an ordinary alarm system has been triggered. Due to methods of construction and utilization, electrical power and ordinary telephone lines in such areas cannot be relied on for security applications. Here emergency radio communications systems with backup battery power supplies are very valuable. These can be arranged to transmit a repeating recorded message upon the triggering of an alarm, to be heard by the police who monitor a particular CB channel in many locations. A good transceiver can be used to summon help for noncrime-related emergencies as well and is generally more reliable than the telephone, especially in rural areas. It is important that the antennas and cables used for emergency transmitters be concealed or made as inaccessible as possible, with sensors to trigger the alarm should they be approached.

Burglars usually steer clear of farm and ranch homes that seem to be occupied for one very good reason. The firearms laws in most such places are sensibly permissive, and most residents have good defensive weapons and know how to use them. In such areas, it is wise to schedule work and absences so that at least one person capable of defending the property is at home at any given time. If no one is at home, make it look like someone is.

The problem of having more than one urban or suburban residence can be minimized by keeping important valuables in the home that you spend the most time in. All residences should be well secured and made to appear occupied when they are not; good alarm systems are a must and central station services are advisable. If you have a vacation home that you use for only a few months in the summer, be sure to strip it of all valuables when you leave, and if you can, rent it out for the time you are away. This does not guarantee its security, but at least you will be compensated for its use and have some choice in who uses it. Happy vacations!

10

INSURANCE

Because automobile liability insurance is generally required by law, there are few people who do not have insurance of one form or another. Based on your circumstances, some forms of insurance are of greater value to you than others. Theft coverage is usually obtained under a "homeowner's" or "tenant's" policy, and what is discussed here pertains mainly to this.

The subject of insurance is not one known for its fascinating excitement. Plodding through fine print, looking for pitfalls with a magnifying glass and a dictionary is far removed from the adventure of a jungle safari, but as with the leopard waiting to spring, your ignorance of their presence and danger can do you harm when you least expect it. A novice in the jungle is at risk. You must ask yourself if you are knowledgeable enough to enter, and if you really need to in the first place. An insurance policy is often a boring perplexity of obscure verbiage. Signing it and paying several hundred dollars without careful evaluation is like buying a used car without looking under the hood. If you have or are considering getting any form of insurance against theft, the following is essential information to you.

PROTECTION FOR THE CARELESS

If you do nothing else, you should get the best insurance policy against burglary that money can buy. If, however, you intend to take the advice and utilize the information given in the last nine chapters, your chances of benefitting from such insurance are very small.

Insurance is a gamble, and unfortunately it subverts taking responsibility for your security. Homeowners who spend little money or effort on the protection of their property are the most likely to collect on an insurance policy. People who make their homes as thiefproof as possible are far less likely to ever file a claim, yet must pay high premiums to support the claims of the careless, as well as the insurance company's profit. It is a system that penalizes responsibility and rewards apathy and negligence. Small discounts on policy premiums are granted by many companies if you have an alarm system of acceptable quality, but to give full credit for the actual reduction in risk these discounts would have to be much larger than they actually are. Premiums are, of course, higher in high-risk areas and increase when risk does, but how often does an insurance company surprise you with a cut in premiums due to a drop in local crime rates or some other liability reduction?

THE ODDS ARE AGAINST YOU

Insurance is a very profitable business. As long as people feel impotent and fearful regarding their ability to protect themselves and their property, and society continues to tolerate crime, it will continue to be a legal way of exploiting the ills of civilization.

As a business, the primary purpose of an insurance company is to make money. They offer a service that you may or may not choose to purchase, and price it so that on the average their income exceeds their outflow as much as possible. This makes it inevitable that the typical policyholder, over many years, will spend far more on premiums than he will collect in claims. For the atypical policyholder—the one who invests in good security to make sure he is very unlikely to ever need insurance—the chances of financial benefit are extremely remote. He is paying a high price for the ever-evasive goal of piece of mind. It makes far more sense to invest in prevention rather than compensation.

Make this test: Total up all the premiums you have paid over ten or more years on homeowner's, tenant's, or even auto insurance. Now total up the value of what has been paid to you in claims. Would you be better off without insurance? Chances are you would, especially if you had sufficient savings to cover the expenses paid for by an occasional claim at the time required. With prudent savings and investment of the money saved on insurance premiums, you most likely would have had enough and more. If, on the other hand, you have benefitted over the long run by having insurance, you may be among the small percentage for whom it is best to have such coverage. Ask yourself though if your claims would have been necessary had you taken greater precautions to prevent the misfortunes, and if such misfortunes are preventable in the future.

Prior to 1983, unreimbursed theft and casualty losses were tax deductible to the extent that the total of such losses exceeded a $100 deductible amount. Effective January 1, 1983, such losses became tax deductible only to the extent that their total exceeds 10 percent of one's adjusted gross income, with a $100 deductible on each loss. This makes good insurance coverage more worthwhile if the chances of such a loss are not minimized. Even with insurance, the loss to the victim due to this change in the tax law is considerable. Before 1983, if a family with a $40,000 gross income sustained a $20,000 burglary loss, with an insurance settlement of $16,000 their net loss would be $4,000, yielding a $3,900 tax deduction. Now, the same circumstances yield no tax deduction at all, as the net loss does not exceed 10 percent of the $40,000 gross income. If uninsured, almost the entire loss would be deductible prior to 1983; now about 80 percent would be. Should this family sustain an uninsured $5,000 burglary loss today, only about 18 percent of the loss would be tax deductible, prior to 1983, 98 percent would have been deductible—a difference of $1,400 in actual loss to a family in the 35 percent tax bracket! Taxpayers would be justified in demanding that every cent of government revenue generated by this tax law revision—in effect a tax increase—be put towards effective measures against serious crime.

POLICY EVALUATION AND INSURANCE COMPANY COPOUTS

Homeowner's and tenant's policies vary considerably in the extent and quality of coverage they offer for theft. Unfortunately, most people do not educate themselves to read and evaluate their policies prior to acceptance. Unbiased expert evaluations are worth having, but the only person who will act 100 percent in your interest is you. Not only should you know how to spot all the loopholes and copouts, you should also know how to find what *isn't* there. A policy may give an impressively long list of property covered and the perils they are insured against, but fail to mention what is *not* covered and the circumstances under which the coverage is void. No print at all can be worse than fine print, and what *is* in print you may be sure has been cleverly worded in the interests of the insurance company. If an insurance agent or company will not permit, or attempts to discourage your careful evaluation of a policy prior to acceptance, walk out and find another one who will honestly respect your right to know what you are paying for. To protect others from such unscrupulous dealings you should also file a complaint with your state department of insurance or other appropriate consumer agency.

The limits of liability can be the most vague and understated section of a policy. You may think that you have excellent coverage on the contents of your home, only to find out after they are stolen that your patio furniture, barbecue, and power mower were not covered. Beware of clauses and fine print that exclude certain personal property from coverage.

Beware of "cash value" settlement terms, as well as depreciation clauses. They result in your being paid only a small percentage of the original cost of a stolen item, even though it was in perfect condition and the cost of replacement is greater than its original cost. Replacement cost coverage is available at a higher price; do not assume you have such coverage unless the policy specifically states it.

It is very unlikely that you will be able to collect for the theft of anything you cannot prove you had. To avoid this, keep sales receipts in your name, and photos, or give the insurance company a property inventory list when you take out the policy. This list should be updated regularly and upon policy renewal. Where value may be questioned, as with antique furniture or collectibles, be sure to get an appraisal.

One common reason for denying payment of a burglary claim is lack of evidence of forced entry to the premises. This means that either the thief was skillful or that you were careless, and in the interests of profit the insurance company assumes the latter. You should not be careless, but unless it can be proven that you were uncommonly negligent to the point of inviting the thief in, you should fight to collect. Some burglary victims have been known to take a crowbar to their own door or window before calling the police so that their insurance company could not evade a settlement. Rather than having to go through all this hassle though, choose your insurance carefully, or better yet, beef up your home security so that you will never need to file a claim in the first place.

The loss of property from a car or other vehicle may be covered either under a homeowner's/tenant's policy or the insurance for the vehicle itself. You are wasting your money if both cover it; keep the better coverage, drop it from the other policy and have your premium reduced accordingly. In thefts of property from vehicles, insurance companies often use the copout of no visible evidence of forced entry, even though many thieves are adept at entering without breaking anything. This practice can lead to a policyholder breaking into his own car when such a theft is discovered. In one outrageous case, an insurance company denied compensation to the owner of a convertible for theft of personal property from it after the stolen car was recovered. The auto had been partially stripped, the part of the convertible top that separates the interior from the luggage compartment had been slashed, and one window had been pulled out, but there were no visible exterior marks to prove forced entry! (Burzynski vs. Norfolk and Dedham Fire Insurance Co., Massachusetts Appellate Court, Western District, No. 211, 357, March 10, 1972.)

Property is generally not covered if it is off your primary home premises and stolen while you are not residing at that place. This can be property that you or a member of your family keeps in a vacation home, dormitory, or camp even when the absence from that location is only for a day or two. The maximum coverage (the most you can collect for a loss) for property away from home is often limited to a small percentage of the total coverage on your home property. In many areas, property away from home is not covered at all, regardless of circumstances.

Items that are not covered by most homeowners' and tenants' policies include property of residents not related to the insured, outdoor furnishings and utensils, pets, rented property, and motorized vehicles (autos, motorcycles, mopeds, power mowers, boats, and so forth). Extended theft coverage (at an extended premium, of course) is usually available for such valuables.

Another copout (wily, aren't they?) is to deny your claim for failure to "promptly" notify the company in writing of the loss, or failure to file a proof of loss form within the time alloted. Send such notification by certified or registered mail directly to the insurance company.

FLOATERS

Floaters are special policies written to cover specific valuables such as jewelry, silverware, or a stamp collection. They also contain a few loopholes for the insurance company's benefit, but on the whole provide much better coverage than a typical homeowner's or tenant's policy. If it is written on an "all risk" basis, a floater will even cover cases of what is called "mysterious disappearance," such as a professional burglary wherein the only evidence of the theft is the disappearance of the property. Items so insured require a bill of sale or an appraisal to determine their value; keep in mind though that the coverage is usually for the actual cash value at the time of loss, not the replacement cost.

AGENT AND INSURANCE COMPANY SELECTION

An insurance agent is a middleman who acts on behalf of one or more insurance companies to sell policies and handle small claims. A broker functions as a middleman between the insured and the company, but does not write policies. In theory the broker acts more on the behalf of the insured, but there is little practical difference because the broker, like the agent, is paid by the insurance company. You must be very careful in selecting an agent or a broker, as many of them are unqualified, incompetent, or worse. You are better off carefully selecting an insurance company on your own and dealing with them directly, but of course this requires that you know specifically what your insurance wants and needs are.

Make no mistake about it, the agent is working for the insurance company and himself, not for you. You do not pay him, although some agents have tried to collect fees from policyholders even when not permitted to by law; again, you must know your rights. Agents select the insurance company for you based on their interests, and in some cases work exclusively for one company. They are paid by commission and receive bonuses based on a low rate of claims (low "loss ratio") from their clients. This affects their choice of company and leads them to discourage policy holders from making small claims by using fear of policy cancellation and higher premiums. The effect on you, the client, is to increase your deductible amount without decreasing your premium accordingly, and your probably winding up with a less than optimum insurance company.

A good insurance agent will meet with you, answer all your questions, and patiently explain in detail the options, benefits, and limitations of available policies. He should not try to pressure you or confuse you, and should be just as friendly and helpful when you file a claim and not try to talk you out of it. Check his reputation with the Better Business Bureau and consumer agencies. He should be a CPCU (Certified Property and Casualty Underwriter) and should have considerable experience in your area of interest; do not go to an auto insurance specialist for a homeowner's policy.

Insurance companies insist on timely payment of premiums and, of course, will acknowledge no claims by anyone who is not specifically covered under the terms of a current policy. That's fine; but do companies always make good on their obligations or do they try to evade them when they feel they can get away with it? The New York State Department of Insurance is one of many such state agencies keeping tabs on the insurance business and hearing complaints. They publish such information as the "Bottom 25" insurance companies with regard to the number of justified complaints received in any given year. Their statistics are informative and should be evaluated by anyone seeking information on the subject or looking for an insurer with integrity. *Consumer Reports* and *Consumers' Research* magazines have published impartial evaluations of insurance companies; check their most recent back issue on the subject at your local library. Other sources of information on insurance companies are *Best's Insurance Reports* and *Best's Insurance Guide.* The Better Business Bureau, as well as your state insurance regulation agency, may be contacted for information regarding the number of complaints filed against various companies; a high rate of complaints is good reason to go elsewhere.

The cost of identical coverage for the same property can vary considerably between insurers; what you should seek is one that keeps premiums low by holding down overhead costs rather than cheating policyholders. Finding a good company takes careful investigation. Be wary of companies that advertise heavily, they may be doing so to combat a bad reputation, and the cost of that advertising adds to the overhead that policyholders wind up paying for one way or another. The best advertisement is a good reputation. It pays to shop around, but first become an educated consumer.

Do not be intimidated by an agent's threats of policy cancellation should you file a claim; should it happen, there are other, perhaps better, sources of coverage available. If a company cancels your policy you are entitled to a pro rata refund (if they cancel with three months remaining of a one year policy, you get back 25 percent of your premium). If, however, you cancel the policy, you do not get a pro rata refund, but one based upon a short-term policy rate.

If you live in a high-crime area or have made many previous claims there is a good chance that you cannot find a private company that will offer you a policy at any price, let alone a reasonable one. If you feel you must have insurance, the Federal Crime Insurance Program is for you. Noncancelable burglary and robbery policies are available in most states for both private and commercial premises. Full information can be obtained by calling 1-800-638-8780 or writing to Federal Crime Insurance, P.O. Box 41033, Bethesda, MD 20814-0436.

Unfortunately, as this book goes to press, it appears probable that the Federal Crime Insurance Program will be terminated in the near future due to Reagan administration budget cuts. The administration claims that this program should be a state responsibility, but strangely, intends to continue the far more costly federal flood insurance program.

The fact is that government, on all levels, is the servant of the people, not vice versa. We must be heard; it is our responsibility to demand that government render criminals harmless to us rather than just put band-aids on the terrible wounds they inflict.

BACKING UP YOUR CLAIM

In order to protect your entitlement to compensation, certain procedures must be followed after a loss is discovered. The following are the most basic requirements in the event of theft:

* The police must be called promptly. Even minor thefts should be reported immediately; if there is no police report, the insurance company may contend that a theft never occurred. Take down the name(s) of the officer(s) responding, their precinct number, the date and time, and the case number assigned to the report. (The insurance company must be given all this information.) Give the police a list of the property taken with all relevant information such as serial numbers and identifying characteristics. If more property is found to be missing later, notify the police and have it added to the report.

* Notify your agent or insurance company by phone *within one day* of the discovery of the theft, if possible—after the police have given you a case number. Tell them you will follow up with a written report. Do so *as soon as possible,* giving all police report information plus a list of the items lost with date of purchase, original cost, and approximate (exact if you can) replacement cost. This is why your inventory list, sales receipts, and photos are valuable; you must be able to prove loss and the value of the loss. Add any incidental costs incurred and state the total loss you are claiming. Send this to your insurance company by certified or registered mail; send a copy to your agent, if you have one, and keep a copy for your records.

* In addition to this notification, you usually must follow up within sixty days of the incident with a "Proof of Loss" form. If your initial written report is complete enough it may suffice, but file this anyway to be on the safe side. If you are told verbally that a proof of loss form is not necessary, get it stated in writing.

* When dealing with an insurance adjuster or investigator, keep in mind that he or she is NOT on your side. These people work for the insurance company and their performance is judged by how much money they save the company on claims, not how fair a settlement you get.

SETTLEMENT DISPUTES

Common disputes relating to theft insurance involve the company's disqualifying your claim, offering an inadequate settlement, or procrastinating settlement. An insurer may attempt to trick you by sending you a check, possibly quite promptly, for an amount less than a fair settlement. DO NOT CASH IT, as doing so may legally waive your rights to further compensation. Patience pays in such cases; take your time and fight. Here is how to do it:

1. Try to resolve your dispute through your agent or the insurance company complaint department. If the problem is an unintentional error, they may help.
2. Go to the top of the insurance company—the president or the director of public relations. If they avoid you, notify them by certified mail. Make sure to state your problem clearly and explain what you expect as a solution.
3. If the insurance company's reply is not to your satisfaction, the next step is your state's department of insurance. Every state in the United States has such a regulatory agency, and this tells you something about the history of the insurance business. They vary in effectiveness from state to state, but often are helpful. Your taxes pay for this service and you are entitled to it, so do not hesitate to use it.
4. Consumer agencies and the Better Business Bureau are worth contacting. They can be very helpful if they have experience with insurance matters or have dealt with a similar complaint against your insurance company in the past.
5. Complain to the media. Insurance companies do not want bad publicity, and consumer assistance features can give them plenty of it. TV, radio, newspapers—use the most effective in your area.
6. Politicians may be of some help if they are not connected with the insurance business. Contact your state senator or representative or your congressperson.
7. In insurance matters, arbitration and the use of lawyers are time consuming and costly. Only resort to these if all else fails and you believe you can win. If you employ a lawyer, only use someone who will accept your case on a contingency basis; this means that the lawyer's fee is a certain percentage of your settlement and if you get nothing, neither does your lawyer.

The Best Insurance
PREVENTION

EPILOGUE

There is no question that some expense and continuing inconvenience is involved in utilizing the security measures given in this book. Unfortunately, we are all paying the price of a society rife with antisocial behavior, and the price can be much higher if we do not invest in prevention and personal protection before we become victims. Don't wait to learn the hard way that it *CAN* happen to you.

This book has been written in the hope that most of us will live to see the day when such books are no longer needed. To hate the criminals among us is to fall victim to them on a spiritual level. Our highest priorities should be to humanely isolate them from constructive society, and to dedicate ourselves to the solutions of the mental and social problems and physical conditions that have contributed to their degradation. When at last we can all fully trust one another, then we will have peace on earth.

INDEX

A

L

M

N

O

P

R